Water Potential Relations in Soil Microbiology

SSSA Special Publication Number 9

Proceedings of a symposium sponsored by Divisions S-1 and S-3 of the Soil Science Society of America. The papers were presented during the annual meetings in Chicago, Illinois, December 3–8, 1978.

Organizing Committee
L. F. Elliott, Chairman
R. I. Papendick
R. E. Wildung

Editorial Committee
J. F. Parr, Chairman
W. R. Gardner
L. F. Elliott

Publications Coordinator
Matthias Stelly

Managing Editor
David M. Kral

Assistant Editor
Mary Kay Cousin

1981
Published by the
SOIL SCIENCE SOCIETY OF AMERICA
677 South Segoe Road
Madison, Wisconsin 53711

Soil Science Society of America
677 South Segoe Road, Madison, Wisconsin 53711 USA

Library of Congress Catalog Card Number: 80-69528
Standard Book Number: 0-89118-767-7

Printed in the United States of America

Contents

Foreword

The number, kind, and activity of microorganisms within the soil affect most basic soil properties. They can, through their effect on soil organic matter, beneficially influence the water holding characteristics, cation exchange capacity, and tilth of soils. Microorganisms decompose plant residues, making available or immobilizing nutrients for crops. Microorganisms oxidize ammonium to nitrate, the mobile form of nitrogen in the profile; fix nitrogen from the atmosphere; and aid in the solubilization of mineral nutrients. Microorganisms living within the soil can also be a source of pathogenic plant diseases.

Even though soil microorganisms play an important role in the soil, their response to the soil environment is not well understood. Obviously, one important environmental factor influencing microbial growth is soil water. This publication *Water Potential Relations in Soil Microbiology* summarizes the most recent research in this important area. It will help in understanding microbial-soil environmental relationships as well as aid in managing soil and crops for human benefit. It will be valuable to the researcher, teacher, manager, and student alike.

The Soil Science Society of America is pleased to present this special publication from the proceedings of a 1978 symposium. The authors are outstanding soil scientists and microbiologists. We are indebted to the symposium organizers, authors, and editors for contributing their time and efforts to the preparation of this publication.

W. E. Larson
President, SSSA

Preface

This special publication presents the ideas and observations of 10 outstanding scientists on the concept and use of water potential applied to soil microbiology and biochemistry, plant pathology, and the microbial ecology of soils. The five papers in this publication were presented at a symposium sponsored by Divisions S-1 and S-3 of the Soil Science Society of America at their annual meetings in Chicago, December 3 through 8, 1978. The organizing committee consisted of L. F. Elliott, R. I. Papendick, and R. E. Wildung, Agricultural Research, Science and Education Administration, U.S. Department of Agriculture.

Objectives of the symposium were to bring together present knowledge of the water potential concept and its application to basic and applied research in soil microbiology and biochemistry, and to project future research needs and applications. The papers cover the theory and measurement of water potential, water potential's effect on the growth, activity, and survival of soil microorganisms, including plant pathogens; and water potential as a selective factor in the microbial ecology of soils.

Water potential is a fundamental concept that can be used to quantify the energy status of water in soil, crop residues, municipal wastes, plants, seeds, and microorganisms. The concept provides a convenient way of applying thermodynamic terminology to the water relations of soil-plant-microbial systems. The water relations of both plants and microorganisms are best described in terms of water's potential energy (i.e. the water potential) because they must expend work or energy to obtain water from their immediate surroundings.

The concept of water potential has been widely accepted by soil scientists in recent years, however many continue to use such soil moisture constants as saturation capacity, water-holding capacity, field capacity, and permanent wilting point. In studies where water is a variable, these parameters have little value in predicting and interpreting the response of microorganisms in soils and crop residues, or in other natural systems involving organic wastes. A strong consensus expressed in this publication is that where water relations of microorganisms are investigated, both the water content and water potential should be reported. This would allow for valid comparisons of data obtained from different investigations involving different soils and experimental conditions. Often such comparisons are virtually impossible because inappropriate constants or indices were used to express the soil water regime.

This publication provides the most complete review to date of the concept of water potential and its importance in studying the water relations of soil microorganisms. It should serve as a basic text and reference for years to come. It is hoped that the publication will provide a better understanding of the concept, its applications, and the methods and techniques available for proper measurement.

J. F. Parr
W. R. Gardner
L. F. Elliott
Editors

CHAPTER 1

Theory and Measurement of Water Potential[1]

R. I. PAPENDICK AND G. S. CAMPBELL[2]

INTRODUCTION

Water flow to plant roots and to microbial activity sites occurs along free energy gradients. The physiological response of living cells to water stress is closely related to the free energy of the water in the cells. Water potential is a fundamental concept now widely accepted for quantifying the energy state of water in soil, organic materials, plants, seeds, and microorganisms. Water potential is the free energy of water in a system, relative to the free energy of a reference pool of pure, free water. Alternatively, it is a measure of the potential energy (per unit mass or volume) of water at a point in a system relative to the potential energy of pure, free water.

Water in unsaturated soil and organic materials, plants, and in living cells is normally subject to certain forces that lower its potential energy relative to free water in the reference state. Since pure, free water is usually assigned zero water potential, the potential energy of water in these systems is almost always negative. In any system, the tendency is toward energy equilibrium. Thus, water flow is spontaneous from high to low potentials, and the availability of water for physiological processes decreases as the potential is lowered.

Components of Water Potential

The total water potential may be expressed as a sum of components identifiable with the forces that retain or act on the water and affect its energy state. These components are: i) the osmotic potential (ψ_π, always negative) due to solutes in the water; ii) the matric potential (ψ_m, always negative), which includes both adsorption and capillary effects of the

[1] Contribution from USDA-SEA AR, Western Region, and the College of Agric. Res. Ctr., Washington State Univ., Pullman, 99164.

[2] Soil scientist and research leader, USDA-SEA-AR, Pullman, Wash., and professor of Soils, Dep. of Agronomy and Soils, Washington State Univ., Pullman, 99164.

Water Potential Relations in Soil Microbiology.

solid phase; iii) the gravitational potential (ψ_g, negative or positive, depending on the reference level), proportional to elevation differences from the reference; iv) the pressure potential (ψ_p, negative or positive, depending on the reference pressure), resulting from external gas or hydraulic pressure applied to the water; and v) overburden potential (ψ_Ω, always positive), caused by weight from overlying matter on water present in a nonrigid porous body. These components are summed to give the total water potential:

$$\psi = \psi_\pi + \psi_m + \psi_g - \psi_p + \psi_\Omega. \quad [1]$$

In most soil and plant systems, matric and osmotic components contribute more significantly to the water potential than the others and exert a greater effect on water flow and availability for physiological processes. The matric potential is the largest component in most unsaturated soils and in crop residues or other organic materials. The osmotic potential is significant in saline soils and, in some cases, with certain crop residues and soils amended with organic wastes or fertilizer.

In living plant cells, including those of microorganisms, the water potential is given by

$$\psi = \psi_\pi + \psi_p \quad [2]$$

The pressure potential in the intact cell is due to turgor. Water in and along cell walls and in the xylem of plant tissue is usually nearly solute free and is, therefore, subject mainly to matric and pressure potentials of the intra- and intercellular wall surfaces.

Differences in water potential across minute distances of microns or less, such as the thickness of a cell wall, can exist only if resistances to flow are extremely high. For this reason, the water potential of a microorganism cell in soil is likely to be in near equilibrium with that of its immediate environment, even though components of the water potential may differ substantially over short distances.

Relationships with Thermodynamic Quantities (Expressions)

The water potential concept provides a convenient means for applying unified thermodynamic terminology to water relations of soil-plant-microbiological systems. By definition, water potential is the difference between the chemical potential of the water (μ_ω) in a particular system and that of pure free water (μ_ω°) at the same temperature, divided by the partial molal volume of water V_ω (1.8×10^{-5} m^3/mole at 4 C) (Slatyer, 1967):

$$\psi = (\mu_\omega - \mu_\omega^\circ)/V_\omega. \quad [3]$$

Since chemical potential is expressed in energy units, water potential is dimensionally equivalent to pressure (Joule m^{-3} = Newton m^{-2}). The bar (= 10^5 N m^{-2}) is commonly used as the standard unit though the Pascal (= 1 N m^{-2}) is the accepted SI unit of pressure. Table 1 gives conversions from

Table 1. Conversion of water potential in bars to other units.

Unit	Value equivalent − 1 bar water potential
ergs/cm^3	-1×10^6
dynes/cm^2	-1×10^6
joules/kg	-1×10^2
cm H_2O	1,020
mm H_g	750
atmosphere	−0.978
newtons/m^2 (pascal)	-1×10^5

bars to other units found in the literature. The bar is generally the most convenient unit for describing soil microbiological phenomena (Griffin, 1972).

The following thermodynamic expressions relate to water potential for equilibrium conditions between the liquid and vapor phases and for water in aqueous solutions:

$$\psi = \frac{RT}{V_\omega} \ln h = \frac{RT}{V_\omega} \ln a_\omega = \frac{RT}{V_\omega} \ln \phi N_\omega. \quad [4]$$

In Eq. [4],

R = universal gas constant (8.31×10^{-5} m^3 bar $mole^{-1}$ K^{-1})

T = absolute temperature (K)

h = equilibrium relative humidity (a fraction)

a_ω = water activity

ϕ = activity coefficient

N_ω = mole fraction of water

For water potentials in the plant growth range, an approximate relationship between water potential in bars and relative humidity at 20 C is $\psi \simeq 1{,}350$ (h-1). A change in relative humidity of 0.01 is therefore equivalent to a change in water potential of about 14 bars. Several values of water potential and their corresponding relative humidities are given in Table 2. Growth of plants and microorganisms often respond to changes in water potential of several bars or less, sometimes even to fractions of a bar (Cook and Papendick, 1972, 1978). This corresponds to relative humidity changes of 0.001 or less. Microbial activity has been reported at humidities below 0.7 (Scott, 1957), which is equivalent to a soil water potential of about −500 bars. Where ambient humidity is above 0.7, it may be possible to measure the activity of certain microorganisms even in air dry soil.

WATER RETENTION BY POROUS MATERIALS

Water in soil and other porous media is retained largely by matric forces in pores and interconnecting pore necks, as lenses at points of contact between mineral and/or organic particles, and as films on particle surfaces. At saturation water content, all of the pores are filled with liquid water and the matric potential, ψ_m, is zero. As water is removed by root absorption, evaporation, or soil drainage, the largest pores empty first because water at particle contact points and on particle surfaces is

Table 2. Relationship between water potential, relative humidity, and effective pore diameter in soil at 20 C.

Water potential bars	Relative humidity	Effective pore diam
		μm
−0.001	1.0000	2,908
−0.002	1.0000	1,454
−0.005	1.0000	582
−0.01	1.0000	291
−0.02	1.0000	145
−0.05	1.0000	58.2
−0.1	0.9999	29.1
−0.2	0.9999	14.5
−0.5	0.9996	5.82
−1	0.9993	2.91
−2	0.9985	1.45
−5	0.9963	0.582
−10	0.9926	0.291
−20	0.9853	0.145
−50	0.9637	0.058
−100	0.9286	0.029
−200	0.8624	--
−500	0.6906	--
−1,000	0.4769	--
−2,000	0.2274	--
−5,000	0.0247	--

held less tightly than in the smaller pores. That is, the matric potential is lower in the small pores than in the large ones. Thus, the water content of the soil or other porous media at a given matric potential depends on the total pore volume that is continuously connected (can be water-filled) and on the size distribution of the pores. These in turn are related to the size of the constituent particles. Coarse-textured soils low in organic matter tend to range in porosity from 0.3 to 0.4, whereas the porosities of clay and organic soils may exceed 0.6.

Moisture Characteristic Curves

Clay soils generally retain larger amounts of water at all matric potentials than loams, which in turn have greater retentive capacity than sands. Typical water potential-water content relationships for sand, silt, and clay soils are shown in Fig. 1. These are called moisture characteristic or moisture release curves. Most agricultural soils have moisture characteristics falling between the sand and clay extremes in Fig. 1. Increasing the organic matter content of a particular soil type generally shifts the retention curve in Fig. 1 to the right. Undecomposed crop residues have a much higher affinity or sorptive capacity for water on a dry mass basis than soils (Fig. 2). The capacity of residues to retain precipitation in the field is appreciable. For example, a surface cover of dry wheat (*Triticum aestivum* L.) residues at 12 mt/ha would absorb about 0.5 cm of rain (saturation) and would prevent much of this water from entering the soil.

At high matric potentials (between 0 and −1 bar), the amount of water retained by soil depends on capillary effects and is therefore strongly influenced by soil structure. At lower potentials, the effect of structure

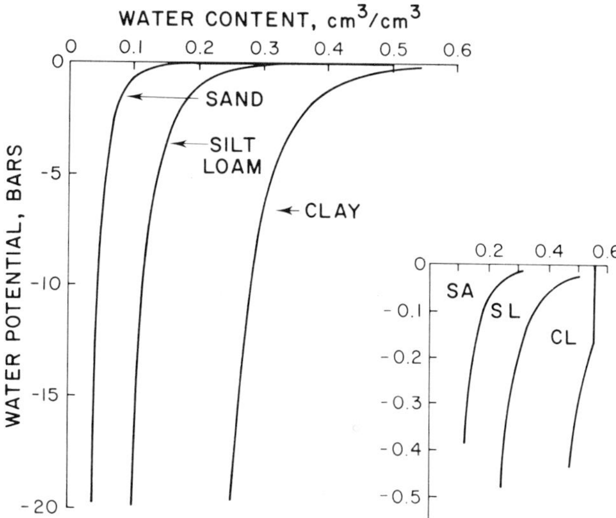

Fig. 1. Water potential-water content relationship (moisture characteristic curves for desorption) for a sandy, silt loam, and clay soil. An expanded scale is used to show water content in the potential range down to −0.4 to −0.5 bars.

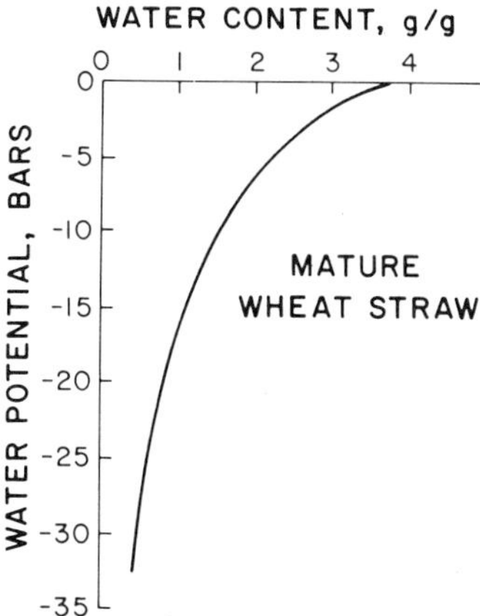

Fig. 2. Water content-water potential relationship for mature, undecomposed winter wheat straw.

is much less pronounced (Campbell and Gardner, 1971) and the shape of the characteristic curve depends more on the soil texture and its specific surface. Caution must be exercised when attempting to extrapolate laboratory measurements of matric potential, especially for wet soils, to field situations; but disturbance of soils at potentials below about −1 bar is likely to have little effect.

The soil moisture characteristic curves of Fig. 1 were obtained by desorption starting with initially saturated soil and drying it until a desired matric potential was reached. If this same water content were reached by sorption of water, beginning with dry soil, the water content for each potential would be lower than for desorption. In other words, the relation between matric potential and water content is not unique, but depends on whether the curve was obtained by wetting or drying the soil. This phenomenon is commonly referred to as hysteresis and is attributed to several causes including effects from nonuniformity of pores, differences in radius of curvature between advancing and receding menisci, effects of entrapped air, and differential changes in soil structure during absorption and desorption (Baver et al., 1972). The hysteresis effect is more pronounced with fine-textured than coarse-textured soils.

Theory has not been developed to mathematically predict the soil moisture characteristic curve from basic soil properties. However, the empirical equation (Gardner et al., 1970; Hillel, 1971)

$$\psi = a\theta^{-b} \quad [5]$$

appears to describe the moisture characteristic for porous materials including some crop residues such as wheat straw (D. D. Myrold, 1979. Effects of water potential and residue placement on the rate of wheat straw decomposition. M.S. thesis. Washington State Univ., Pullman) within a limited range of matric potentials. In Eq. [5], θ is the water content and a and b are constants for a given soil or other material. The a and b values are obtained by taking the logarithm of both sides of Eq. [5] to obtain $\ln \psi = \ln a - b \ln \theta$ (ψ is taken as positive here). This is a linear equation of the form $y = mx + c$, with $y = \ln \psi$, $x = \ln \theta$, $m = -b$, and $c = \ln a$. A linear regression of $\ln \psi$ and $\ln \theta$ will therefore give $b = -m$ and $a = \exp c$.

Equation [5] makes interpolation between data points on a characteristic curve feasible. The points are plotted on log-log paper and a straight line drawn between them.

Relationship of Soil Water Potential to Some Moisture Constants

Sometimes researchers express soil water in terms of so-called soil moisture constants such as "saturation capacity" (SC), "water-holding capacity" (WHC), and "field capacity" (FC), as upper limit references, and "permanent wilting percentage" (PWP) as a lower limit reference. Moisture levels are then established relative to these limits (e.g., 3/4 FC, 60% WHC). Most workers take SC and WHC as the water content at saturation, but some use WHC to designate water contents ranging from FC to saturation. The accepted definition of FC is "the water content of a soil profile, usually the rooting zone, which has been thoroughly wetted by irrigation or rainfall, and after the subsequent rate of drainage out of the profile has become negligibly small" (Baver et al., 1972). Permanent wilting point is the soil water content at the time leaves of indicator plants growing in small containers of soil become wilted and do not recover turgor overnight.

If SC or WHC are taken as saturation (all pores filled with water), then the matric potential is zero at these water contents. The water potential at field capacity is always arbitrary, depending on one's definition of a negligibly small rate of drainage. For crop conditions, where the negligible drainage rate is assumed to be about a tenth of the evapotranspiration rate, the water potential of a uniform soil profile at FC approaches −0.1 bar for sandy soils and −0.3 bar for loams and silt loams. The presence of coarse layers in the soil may increase these values substantially (Clothier et al., 1977). The permanent wilting point is around −15 bars, which is approximately equal to the tissue osmotic potential of the indicator plants used in its determination. Under field conditions, plants can often dry a soil to potentials well below −15 bars. We have measured root zone water potentials as low as −45 bars under winter wheat (Papendick et al., 1971) and −70 bars under sagebrush (*Artemisia tridentata* Nutt.) (Campbell and Harris, 1977). The soil water potential which causes wilting therefore depends on plant, soil, and atmospheric factors, and is not a constant. In practice, however, the water content at −15 bars can often be used as a lower limit of available water because the characteristic curve is so steep at these potentials that several bars of uncertainty in ψ_T results in only a small uncertainty in PWP.

The soil moisture constants are useful for designing irrigation systems and calculating water budgets for crops, but they have little value in studies of the response of microorganisms to water. For such studies, both the water contents and the water potentials for various systems should be reported.

SOIL MATRIC POTENTIAL, SIZE OF WATER-FILLED PORES, AND FILM THICKNESS

Most soils are comprised of a wide range of particle sizes, from 2 mm equivalent diam coarse sand to <0.002 mm clay. The individual grains are packed or arranged in different ways which result in a broad distribution or range of pore sizes. Moreover, in a well-structured soil, small pores may exist in individual aggregates or crumbs with larger pores in the spaces between the aggregates. A narrow pore size distribution (pores essentially of the same size) would occur only in a system of particles of essentially the same shape and size (e.g., spheres with a common diam). The capillary rise equation relates the matric potential of water in a capillary to the radius of curvature r (μm) of the meniscus: $\psi_m = 2\,\sigma/r$, where σ is the surface tension (72.7 g sec^2 or 0.727 bar μm at 20 C). Table 2 shows equivalent pore diam for a range of water potentials, calculated using the capillary rise equation. The calculation of pore size is only approximate because surface tension is not precisely known.

The capillary rise equation is often used with the characteristic curve to estimate the pore size distribution in soil. A decrease in matric potential from ψ_{m1} to ψ_{m2} will result in a release of water from pores of an effective diam ranging from d_1 to d_2. Since the relationship assumes that the contact angle between water and soil solids is zero and that pores are cylindrical, the approximation is reasonable only at high matric potentials (e.g., >1 bar).

The pore size distributions for the soils of Fig. 1 calculated from the capillary rise equation are presented in Fig. 3 and 4. The curves reveal large differences in classes of pore sizes for the three soil types. Forty percent of the sand pores are larger than 100 μm, while less than 10% of the

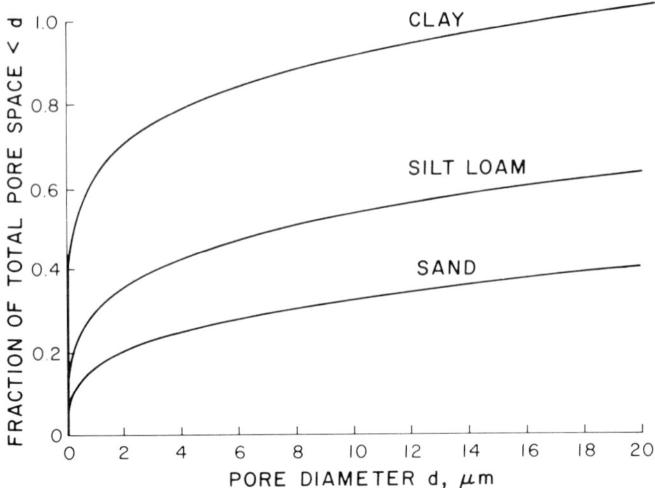

Fig. 3. Fraction of pore sizes having effective pore diameters less than d micrometers for the range of 0 to 20 μm for the sand, silt loam, and clay soils of Fig. 1.

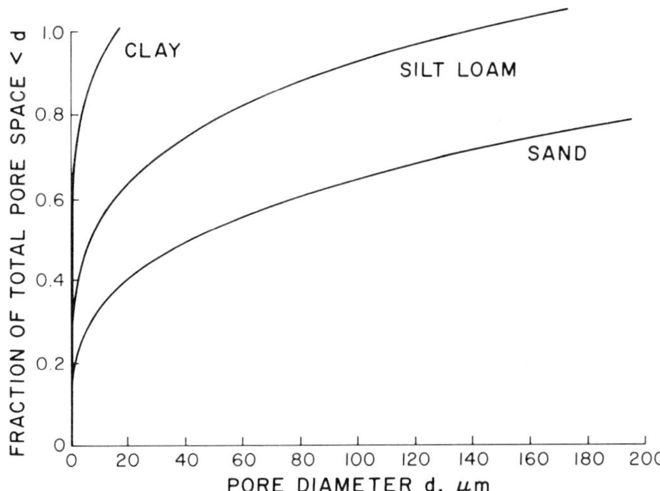

Fig. 4. Fraction of pore sizes having effective diameters of less than d micrometers for the range of 0 to 200 μm for the sand, silt loam, and clay soils of Fig. 1.

silt loam pores and none of the clay pores are this large. Over 60% of the clay pores are smaller than 1 μm, while only 30% of the silt loam pores and less than 20% of the sand pores are this small.

When a saturated soil is allowed to drain, a certain amount of tension or pull must be applied to the water before the largest pores will begin to empty. The matric potential at which the pores of a saturated soil first begin to drain is called the air entry potential. Air entry in soil occurs at high potentials, but varies with the soil type. For the soils of Fig. 1, air entry potentials were −0.007, −0.02, and −0.17 bar for the sand, silt loam, and clay, respectively. Thus, the equivalent diam of the maximum size water-filled pore at the air entry value is 442, 146, and 17 μm for these soils in the same order.

Along with Table 2, the curves in Fig. 3 and 4 can be interpreted to illustrate two important points: i) soils vary greatly in both pore size distribution and the pore volume that remains water-filled over a range of matric potentials; and ii) in the range of matric potentials that water is most readily available to plants, the percent of water-filled pore space, for sand and loam soils, is relatively constant and in the range of 30 to 50%.

At water potentials below −1 bar, the water content-water potential relationship of soil is dominated by surface adsorption effects. According to Richards (1960), when thickness of adsorbed water films is reduced to 6 or 8 molecular layers of water, plants are virtually unable to extract the water and growth ceases. He calculated that for a clay soil with a specific surface of 220 m^2/g and a water content of 0.22 m^3/m^3, the water, if spread evenly, would cover this surface with a thickness of only 0.8 nm.

Effects on Movement of Organisms in Soil

The physical configuration of water in soil which relates directly to water potential may influence the motility of certain microorganisms as well as diffusion of gases, nutrients, and exudates to and away from sites of biological activity. This is especially true of microorganisms that lack a hyphal system to bridge air spaces. The movement of nematodes, motile bacteria, and aquatic phycomycetes may be limited if the water-filled pores or pore necks in the soil are too small to permit their passage (Griffin and Quail, 1968; Stolzy et al., 1965; Wallace, 1958). Different organisms, presumably because of their size characteristics, may require different water-filled pore size ranges for optimum movement. For example, for maximum movement and dispersion in soil, fungi of the genus *Phytophtora* require water-filled pores at least 40 to 60 μm in equivalent diam (Stolzy et al., 1965). Larvae of the beet eelworm (*Heterodera schachtii*) moved most rapidly if the water-filled pore diam was 30 to 60 μm (Wallace, 1958). The bacteria *Pseudomonas aeruginosa* requires water-filled pores of 1 to 1.5 μm in diam or larger to move readily in soil (Griffin and Quail, 1968).

Wallace (1958) reported that the extent of microbial movement in soil is not only a function of water potential, which controls the size of water-filled pores, but also of particle size and packing which controls the pore numbers of a specific size range. He showed that the movement of *H. schachtii* larvae in soil was optimal at matric potentials of −0.25 to −0.04 bar where the particle size was 150 to 250 μm, because these conditions provided the maximum number of water-filled pores of ideal size for the organisms.

The spread or dispersion of a specific organism in soil at a given water potential may vary considerably with soil type because of differences in numbers of pore sizes over the range favorable for its movement. However, pore size is not the only restriction to movement of organisms. The air-water interface itself can hold the organism down like a rubber membrane.

Effects on Diffusion of Gases in Soil

At high matric potentials, water in soil pores may limit exchange of O_2 and CO_2 around centers of microbiological or root activity. The diffusion coefficient of O_2 in air has been reported as 0.189 cm^2/sec (Letey et al., 1967), this being about 1.25 times that for CO_2. In water, the diffusion rate of both gases is about 1/10,000 that in air (Baver et al., 1972) (e.g., the reported diffusion coefficient of oxygen in water is 2.56×10^{-5} cm^2/sec). Therefore, with water present in soil, the effective area for diffusive flow for either gas is reduced in proportion to the pore space occupied by water. For a wide range of porosities, the apparent diffusion coefficient D of a gas in a dry nonreactive porous solid is $D = D_0S^m$ (Currie, 1965), where D_0 is the diffusion coefficient of the gas in air and S is the air-filled porosity (space not occupied by solid matter). With water present, the air-filled porosity becomes $S = \theta_s - \theta$, where θ_s is the water content at saturation. The value of m varies with different soils over a range of about 1 to 3 (Currie, 1965). A typical value may be near 2.

If the apparent diffusion coefficient of a gas in moist soil is proportional to the square of the gas-filled porosity (m = 2), slight changes in the water content of soils near saturation would be expected to markedly influence the gas diffusion rate. In soils that are wet enough to limit gas diffusion, matric potential is not likely to have much influence on microbial activity since potentials are near zero. Since aeration is related directly to soil water content for all textural classes of soil, water content is probably the best moisture variable to use in studying the effects of limited aeration. Figure 5 shows that, if matric potential were related to the diffusion coefficient for the soils in Fig. 1, large differences in D would be predicted for these soils at a given matric potential. At −0.5 bar matric potential, the apparent diffusion coefficient for the sand is about 10 times that for the clay. With decreasing potential, the change in D is more gradual for all soils. The sand and loam are much less restrictive to aeration than the clay over the range of soil matric potentials where most plants grow. The influence of matric potential on soil aeration may help explain why matric potential effects on certain biological processes are not the same for different soil types.

Effects on Solute Diffusion

A consistent feature in comparisons of microbial activity under osmotic- and matric-induced water stress is that activity is reduced substantially more by lowering the matric potential than by lowering osmotic potential (Adebeyo and Harris, 1971; Cook et al., 1972; Griffin, 1972).

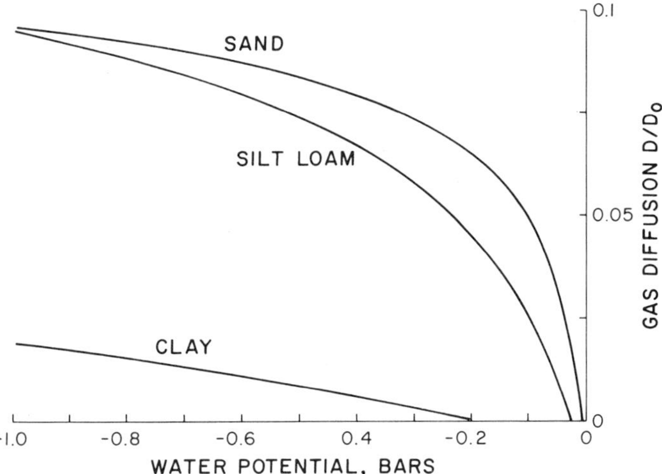

Fig. 5. The relative gas diffusion coefficient D/D_0 for water potentials ranging from 0 to -1.0 bar for the sand, silt loam, and clay soils of Fig. 1. D is the apparent diffusion coefficient in soils, D_0 is the diffusion coefficient of the gas in air.

Various authors have suggested that solute diffusion to and from microorganisms is more severely limited to systems with matric-induced stress than in osmotic systems with higher water content. Thus, reduced activity results more from an indirect effect on substrate availability and transport of metabolic products rather than a direct effect of water potential on microorganisms. This is especially apparent in the nitrification process where small reductions in water potential (< -0.2 bar) can substantially reduce the rate of nitrification (Sabey, 1969). It is unlikely that the metabolic processes of the microorganisms are influenced directly by such small changes in water potential.

Solutes could be transported to and away from microorganisms by mass flow or diffusion. Since the water content of the microorganism stays almost constant, the major contribution of mass flow to solute uptake would come from solutes moving with the water that flows past the microorganism in the normal course of water infiltration and redistribution in the soil. If one assumes that the nutrients within a cylinder twice the diam of a bacterium are extracted from water flowing past, simple calculations indicate that the supply of nutrients from this source would be about two orders of magnitude smaller than that supplied from diffusion. We conclude, therefore, that mass flow is generally unimportant in supplying nutrients to bacteria. Mass flow is important, however, in replenishing nutrients in the bulk soil solution.

The rate of diffusion J of substrate in water to a microorganism is given by

$$J = (c_o - c_b)/r \qquad [6]$$

where c_o and c_b are substrate concentrations in the bulk soil and at the microorganism surface, and r is the resistance to diffusion. If the microorganism is assumed to be a sphere whose radius is small compared to the length of the flow path, then the resistance is (Campbell, 1977)

$$r = s/D \tag{7}$$

where s is the diam of the microorganism and D is the diffusion coefficient for the substrate in the soil water. The diffusion coefficient is a function of the physical and chemical properties of the soil. Olsen and Kemper (1968) give D as

$$D = D_o(L/L_e)^2\theta\alpha\gamma \tag{8}$$

where D_o is the diffusion coefficient (m^2s^{-1}) of the substrate in water, θ is the volumetric soil water content, α is a factor to account for the increased drag due to the higher viscosity of water near particle surfaces, and γ is a factor to account for electrostatic forces on charged diffusing substances. The ratio, L/L_e, accounts for effects of tortuosity on the diffusion coefficient. The macroscopic length of the flow path is L, and the actual path taken by a diffusing solute particle is L_e. The factor is squared because this effect acts both on the distance a particle must travel and on the concentration gradient which is the driving force for diffusion. Following Brooks and Corey (1966), we assume that L/L_e is a linear function of water content, so that $(L/L_e)^2 = k\theta^2$, where k is a constant. If we assume that the increased drag due to electrostatic forces and to increased water viscosity are negligible (an assumption that needs further examination), then

$$D = D_o k\theta^3 \tag{9}$$

and

$$r = s/(D_o k\theta^3). \tag{10}$$

If rates of microbial processes can be assumed to follow a Michaelis-Menten relationship with a specific nutrient concentration at the microbial surface, then the flux of nutrients to the microorganism is given by

$$J = \frac{c_o + K + rJ_m - \{(c_o + K + J_m r)^2 - 4rc_oJ_m\}^{1/2}}{2r} \tag{11}$$

where K is the Michaelis-Menten rate constant, and J_m is the maximum possible process rate when the substrate is not limiting.

An indication of the significance of diffusion on rates of microbial processes was obtained by calculating J for nitrification as a function of water content using Eq. [10] and [11]. Constants were available from the literature (Ardakani et al., 1974; McLaren, 1970). We used a value of 0.05 g $m^{-2}s^{-1}$ for J_m (Ardakani et al., 1974) and 15 mg/kg for K (McLaren,

1970). Other constants were s = 0.5×10^{-6} m, D_0 = 2×10^{-9} m^2s^{-1}, and k = 2.8. The effect of water content on nitrification, based on this model, is shown in Fig. 6. Predicted activity of microorganisms drops rapidly between 20 and 10% soil water content, even when substrate concentrations are high. As substrate concentrations decrease, the water content at which activity is reduced increases.

The water potential at which biological activity is reduced depends, of course, on the functional relationship between water potential and water content as it did with aeration in the previous section. Since diffusion of solutes is directly related to the cross section for flow, rather than the energy status of water in the system, water content is the proper variable to use in comparing reaction rates in systems which are diffusion limited.

Figure 7 gives an indication of the possible confusion which could result if water potential was used as the variable in a diffusion-limited system. In Fig. 7, the data from c = 100 curve of Fig. 6 are replotted as a function of water potential for the three soils in Fig. 1. Diffusion differs appreciably at a given potential for the three different soils. However, if plotted on a water content basis (not shown) diffusion would be similar for the three different soils.

METHODS FOR MEASURING WATER POTENTIAL

Numerous methods are available for measuring water potential. Many have been reviewed elsewhere (Black et al., 1965; Griffin, 1972; Brown and VanHaveren, 1972) and will not be discussed here. Those

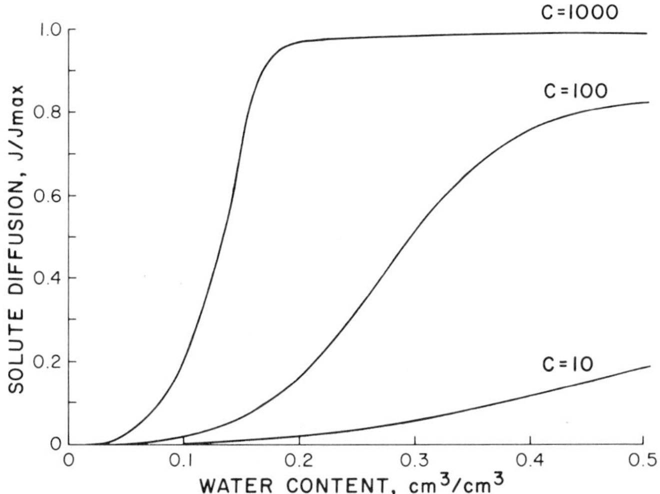

Fig. 6. The relative solute diffusion rate, J/J_{max}, for soil water content ranging from 0 to 0.5 cm^3/cm^3 at three different ambient solute concentrations (C = 10, 100, and 1,000 mg/kg). J is the actual diffusion rate of the solute in soil, and J_{max} is the maximum uptake rate when the ambient solute concentration does not limit uptake.

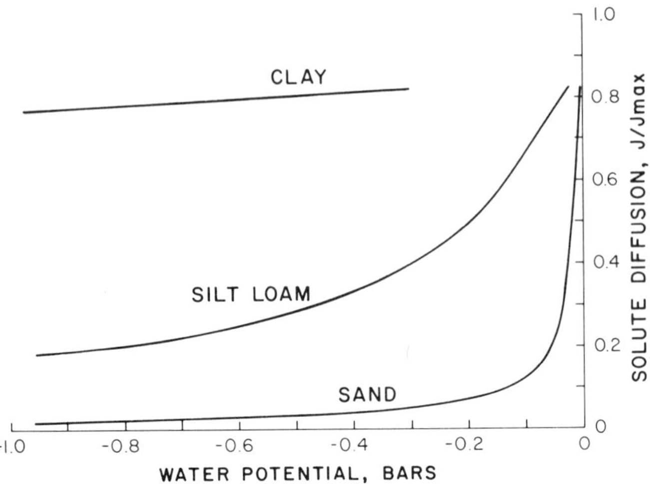

Fig. 7. The relative solute diffusion rate, J/J_{max}, for C = 100 mg/kg as a function of water potential for the sand, silt loam, and clay soils of Fig. 1.

most likely to be useful in soil microbiology are the suction plate (or tensiometer), the pressure plate, and the thermocouple psychrometer.

The suction plate consists of a porous plate that is permeable to water and solutes, but not to air, connected to a water reservoir which can be placed under suction. A soil sample is placed on the plate and the soil solution is equilibrated with the water in the reservoir. At equilibrium, the water in the soil is at the same matric potential as the water in the reservoir. The potential of the water in the reservoir is equal to the pressure on the water, which is determined by the suction applied to it (usually by a hanging water or mercury column). At pressures around -0.8 bar, the water in the reservoir cavitates forming vapor bubbles. Thus, the range of the suction plate is limited to water potentials between 0 and -0.8 bar. The porous plate should have a bubble pressure in excess of 1 bar, so the maximum pore diam should be smaller than 3 μm.

The tensiometer operates in exactly the same way as the suction plate except that the porous filter is constructed as a bulb or cylinder for insertion into a mass or body of soil to equilibrate with the soil water.

The pressure plate uses a principle similar to that of the suction plate, but pressure is applied to the sample to raise its water potential. The potential of the water reservoir is maintained at zero. At equilibrium, the total water potential of the soil sample is zero (equal to that of the reservoir) so the matric potential is the negative of the pneumatic pressure applied to the sample. In using the pressure plate for microbiological studies, it is important to remember that microbial response can depend on the partial pressure of gases in the soil atmosphere. When air pressure is applied, the partial gas pressures change drastically. Gases can be mixed to maintain constant partial pressures of O_2 and N_2, but this is difficult and cumbersome. Use of the pressure plate should, therefore, generally be restricted to determination of moisture characteristic curves for soils. Water potentials can be inferred from water content measurements; however, care must be exercised, since hydraulic conductivities are very low at low water potentials, and equilibrium in the pressure plate can take considerable time or be difficult to achieve. Lack of equilibrium leads to erroneous moisture release curves. Errors in the opposite direction are possible if water evaporates from the sample as a result of air leakage and temperature gradients within the apparatus. The operator must always ensure that samples are properly covered within the pressure plate apparatus. The range of operation for a pressure plate is generally 0 to −15 bars or lower.

Thermocouple psychrometers are frequently used to measure water potential in studies of water relations of microorganisms (Gardner et al., 1972). An advantage is that they can be used to measure potentials throughout the moisture range that may affect microbial growth and activity and measurements can be made in situ. Psychrometers measure the sum of osmotic and matric potential, which is usually the potential of interest in microbial studies. Disadvantages of thermocouple psychrometers are that the equipment tends to be complicated and measurements are strongly influenced by temperature gradients within the apparatus and the cleanliness of the equipment.

The methodology for thermocouple psychrometry has been reviewed elsewhere (Brown and VanHaveren, 1972; Wiebe et al., 1971) and needs only to be presented briefly here. The basic principle is that, if the water in a sample is allowed to equilibrate with water vapor in air in a closed system, the water potential in the liquid and vapor phases will be equal. The water potential of the vapor phase is uniquely related to the relative humidity by Eq. [4]. One simply uses the cooling of a wet bulb thermometer to measure the relative humidity of the air in equilibrium with the sample and then calculates the water potential by Eq. [4]. The temperature depression is found by cooling a thermocouple in a sealed sample chamber using the Peltier effect until moisture condenses on it. The temperature depression of the wet bulb is then measured using the Seebeck effect to provide an output voltage from the thermocouple which is proportional to the wet bulb temperature depression. Problems arise from the narrow range of humidities encountered in living systems (Table 2). For 0.1 bar accuracy in water potential measurement, humidity must be measured to better than one part in 10,000. Two problems need particular attention if measurements are to be made with this accuracy. First, the chambers containing samples must be sufficiently clean and constructed from the proper materials so that a minimum amount of vapor is adsorbed in the chamber and true humidity equilibrium can be achieved in reasonable time. Second, wet bulb depression needs to be measured with sufficient accuracy to determine the humidity. The major problem is that of determining the dry bulb temperature at the sample surface accurately. The wet bulb depression is generally determined by measuring the change in thermocouple output that results from condensing water on the thermocouple. The proper wet bulb depression is the temperature difference between the sample surface and the wet bulb. If the dry thermocouple temperature is different from the sample temperature, the psychrometer reading will be incorrect. A temperature difference of 1 C between the sample and dry thermocouple will result in a measurement error of about 130 bars at room temperature. Obviously, the thermocouple psychrometer sample chamber must be precisely isothermal for measurements to be reliable. For this reason, early workers using psychrometry made measurements in carefully controlled constant temperature baths. More recently, it has been found that sample changers can be used without precise temperature control if temperature fluctuations are minimized. Water baths without temperature control, or enclosures constructed of materials having high thermal conductivity, provide the necessary temperature stability. In situ measurements require either temperature stabilization using high thermal conductivity materials (Campbell and Campbell, 1974; Neuman and Thurtell, 1972) or careful attention to design so that vapor and heat paths are similar (Rawlins and Dalton, 1967).

While we have used both in situ and sample methods in water potential studies, we have found the sample measurements to be the most reliable. Osmotic potentials can be quickly and accurately determined with commercially available sample chambers (Wescor, Inc., Logan, Utah, Model C-52).[3] Two versions of the sample chambers are available, one which is constructed from nylon and the other from aluminum. The aluminum (C-52) unit is far superior to the nylon (C-51) counterpart for reasons of temperature stability mentioned earlier. Water potentials of soil, agar, straw, and other materials can also be determined using the Wescor sample chamber, but the process is much more time consuming than with osmotic samples because samples can only be loaded singly and you have to wait for temperature equilibrium with each sample. We have used the sample changers of Campbell et al. (1966) with good success by simply placing them in a water-filled styrofoam picnic box for temperature control. This allows six samples to equilibrate at one time. This apparatus is also easier to use for samples with water potentials ranging from -50 to $-1{,}000$ bars and where water cannot be condensed using Peltier cooling. In this case, the thermocouple is cooled over -10 bar KCl and then rotated to the dry sample to obtain the wet bulb depression (Campbell and Wilson, 1972).

In using the thermocouple psychrometer to measure soil water potential, you must remember that it measures the sum of matric and osmotic potentials. With most soil samples, however, the osmotic contribution to the total potential is negligible. If it is necessary to know the matric potential when the osmotic contribution is significant, it is usually possible to estimate the osmotic potential of the saturation extract. If the osmotically active component is primarily due to electrolytes, the electrical conductivity (EC) of the saturation extract can be used to infer osmotic potential. An approximate relationship is $\psi_{\pi s} = 0.36\ (\mathrm{EC} \times 10^3)$ where $\mathrm{EC} \times 10^3$ is in mmho/cm, and $\psi_{\pi s}$ is in bars (U.S. Salinity Labora-

[3] Trade names and company names are included for the benefit of the reader and do not imply endorsement or preferential treatment of the product by the USDA.

tory Staff, 1954). Once this value is known, then the osmotic potential at any water content is

$$\psi_{\pi} = \psi_{\pi S} (\theta_S/\theta) \quad [12]$$

where θ_S is the saturation water content.

CONTROL OF WATER POTENTIAL IN MICROBIOLOGICAL AND BIOCHEMICAL STUDIES

The experimental situations where control of water potential is important are too numerous to cover here. Instead, we will explain some general principles that may apply to many situations. In general, experimental designs are based on near equilibrium conditions where the potential is either osmotic or matric. This often involves controlling the total potential of a particular substrate.

If the osmotic potential of soil or agar media is to be controlled, you must know the salt concentration necessary to achieve a given osmotic potential, which is then calculated using a modified van't Hoff relation

$$\psi_{\pi} = \phi\gamma cRT \quad [13]$$

where c is the concentration (moles/kg) of the solute, R is the gas constant (0.0831 bar kg $mole^{-1}K^{-1}$), T is the Kelvin temperature, ϕ is the osmotic coefficient, and γ is the number of osmotically active particles per molecule of solute (i.e., 2 for NaCl). For many studies, osmotic potentials can be computed with sufficient accuracy from Eq. [13] by simply assuming $\phi = 1$. If higher accuracy is required, one must consult a table of osmotic coefficients such as those given by Robinson and Stokes (1965). Table 2 gives osmotic coefficients at 25 C for some commonly used osmotic agents. Available data on temperature dependence of ϕ indicates that its variation is less than ± 2% over the range 0 to 50 C for KCl (Campbell and Gardner, 1971) and NaCl (Lang, 1967). For most purposes it is probably safe to assume that the ϕ values in Table 3 are independent of temperature and that the temperature dependence of π is given by Eq. [13].

When mixtures of solutes are used to control ψ_{π}, the osmotic potentials of the constituents of the mixture are summed to give the total potential (Robinson and Stokes, 1965). Calculations can be verified by measuring the osmotic potential with a thermocouple psychrometer or other type of osmometer.

If the matric potential of soil is to be controlled, we must first consider the range of potentials in a particular study. If the potential is always to be above −0.8 bar, then the suction plate apparatus may be the easiest to use. It is relatively easy to construct and provides control of the water potential and water content without restricting aeration. Samples drier than −0.8 bar can be wetted to the water content required for a given water potential, as determined from moisture release curves. Water

Table 3. Osmotic coefficients of solutions [from Robinson and Stokes (1965)].

M	NaCl	$NaNO_3$	NaH_2PO_4	KCl	KNO_3	KH_2PO_4	$MgSO_4$	Sucrose	LiCl	M	LiCl
0.1	0.932	0.921	0.911	0.927	0.906	0.901	0.606	1.008	0.939	7	1.965
0.2	0.925	0.902	0.884	0.913	0.873	0.868	0.562	1.017	0.939	8	2.143
0.3	0.922	0.890	0.864	0.906	0.851	0.843	0.540	1.024	0.945	9	2.310
0.4	0.920	0.881	0.847	0.902	0.833	0.823	0.529	1.033	0.954	10	2.464
0.5	0.921	0.873	0.832	0.899	0.817	0.805	0.522	1.041	0.963	11	2.607
0.6	0.923	0.867	0.819	0.898	0.802	0.789	0.518	1.050	0.973	12	2.730
0.7	0.926	0.862	0.808	0.897	0.790	0.773	0.517	1.060	0.984	13	2.830
0.8	0.929	0.858	0.798	0.897	0.778	0.760	0.518	1.068	0.995	14	2.915
0.9	0.932	0.854	0.789	0.897	0.767	0.747	0.520	1.079	1.006	15	2.978
1.0	0.936	0.851	0.780	0.897	0.756	0.736	0.525	1.088	1.018	16	3.023
1.2	0.943	0.845	0.765	0.899	0.736	0.716	0.542	1.108	1.041	17	3.044
1.4	0.951	0.839	0.751	0.901	0.718	0.698	0.567	1.129	1.066	18	3.057
1.6	0.962	0.835	0.739	0.904	0.700	0.683	0.597	1.150	1.091	19	3.066
1.8	0.972	0.830	0.729	0.908	0.684	0.669	0.630	1.169	1.116	20	3.063
2.0	0.983	0.826	0.721	0.912	0.669	--	0.666	1.189	1.142	--	--
2.5	1.013	0.817	0.705	0.924	0.631	--	0.780	1.240	1.212	--	--
3.0	1.045	0.810	0.696	0.937	0.602	--	0.922	1.288	1.286	--	--
3.5	1.080	0.804	0.691	0.950	0.577	--	--	1.334	1.366	--	--
4.0	1.116	0.797	0.691	0.965	--	--	--	1.375	1.449	--	--
4.5	1.153	0.792	0.694	0.980	--	--	--	1.414	1.533	--	--
5.0	1.192	0.788	0.699	--	--	--	--	1.450	1.619	--	--
5.5	1.231	0.787	0.706	--	--	--	--	1.482	1.705	--	--
6.0	1.271	0.788	0.713	--	--	--	--	1.511	1.791	--	--

potentials and water contents should be checked on subsamples by using the thermocouple psychrometer and then oven-drying the sample. After the sample has reached a constant water content, the aeration requirement may present some problems. The vapor pressure gradient between the sample and its environment must be small during incubation. This requirement is easily met for samples at potentials within the plant growth range by placing samples in a saturated atmosphere (i.e., a box with wet filter paper around the walls). Diffusion resistance from the sample to the air should be maximized, consistent with providing adequate aeration.

Samples with humidities substantially below 1.0 should be maintained in an atmosphere with a relative humidity near that of the sample. The humidity of the atmosphere can be maintained using osmotic solutions as previously mentioned. The conventional approach is to use a dessicator jar with the salt solution in the bottom. The temperature gradient effects mentioned earlier regarding psychrometers can cause large errors in the control potential with such a system. More reliable control is provided using the agar dish method of Harris et al. (1970). A solute-amended agar gel disk having the appropriate water potential is attached to the inside of a petri dish cover to ensure a short distance between the solution and the sample. Temperature gradient effects are therefore minimized.

Water potential control in dynamic systems is much more difficult to achieve. One system particularly interesting to microbiologists is control of water potential in the rhizosphere. Since gradients of water potential in this region are steep (Papendick and Campbell, 1975), control of bulk soil water potential does not control the potential in the rhizosphere. If it is necessary to control water potential within the rhizosphere, the system of Hsieh et al. (1972) can be used. Plants are grown on a fine screen so that root hairs penetrate the screen and withdraw water from soil beneath; however, roots do not penetrate the screens because of the small openings. Water flow to the roots is controlled by adjusting the soil water tension. The water content within the rhizosphere, and immediately adjacent to it, is measured with a collimated beam of gamma radiation. The water potential is inferred from a moisture release curve for the soil.

SUMMARY

Water potential is a fundamental concept now widely accepted for quantifying the energy state of water in soil, organic materials, plants, seeds, and microorganisms. The concept provides a convenient means for applying unified thermodynamic terminology to water relations of soil-plant-microbiological systems. Of the six components which are usually summed to give total potential in soil and plant systems, the matric, osmotic, and pressure (turgor) potentials are most important in systems of concern to microbiologists. The matric potential of soil and certain organic materials can be inferred from measurement of water content through moisture characteristics or moisture release curves. A simple power-law equation appears to adequately describe the moisture characteristic for many porous materials within a limited, but useful, range of matric potentials.

Soil moisture constants such as "saturation capacity," "water-holding capacity," "field capacity," and "permanent wilting point" have little value in interpreting microbial response to water. In studies of water relations of microorganisms, both the water contents and the water potentials of the various systems should be reported.

The maximum size of water-filled pores in porous materials is inversely related to the matric potential through the capillary rise equation. Thus, matric potential in combination with pore size distribution, which is determined primarily by particle size (textural class) and structural configuration of the matrix, exert a strong influence on organism dispersion in soil by regulating the size and volume of water-filled pores available for their movement.

Diffusion of gases and solutes is directly related to water content for all textural classes of soils as well as other porous materials and not to the energy status of water in the system. Therefore, water content, not water potential, is the proper variable to use in comparing microbially-related reaction rates in systems which are diffusion limited.

Methods most useful for measuring water potential in soil microbiology are the suction plate (or tensiometer) or pressure plate which measures matric potential, and the thermocouple psychrometer which measures the total potential (which for many systems is the sum of the matric and osmotic potentials). An advantage of the psychrometer is that it can be used to measure potentials throughout the moisture range that may affect microbial growth and activity, and measurements can be made in situ as well as on soil, plant, and organic material samples. When using the psychrometer to determine matric potential of saline soils, the osmotic component may be estimated from an approximate relationship with the electrical conductivity of the saturation extract.

Control of water potential in microbiological and biochemical studies is usually based on near equilibrium conditions where the potential is either osmotic or matric. This often involves controlling the total potential of a particular substrate. Osmotic control of soil or agar media is achieved using salt solutions of a given osmotic potential which can be calculated from a modified van't Hoff relationship. For higher accuracy, osmotic coefficients may be used in the calculation. Control of matric potential in soil may be achieved with a suction or tension plate in the wet range (e.g., > -0.8 bar) with a pressure plate apparatus down to -15 to -20 bars, or by wetting the soil to required water content as determined from a moisture release curve. Specialized techniques are available for controlling water potential through control of equilibrium relative humidity. Specialized techniques are also available for control of water potential in dynamic systems, but these are more difficult to use than methods for static systems.

LITERATURE CITED

1. Adebeyo, A. A., and R. F. Harris. 1971. Fungal growth responses to osmotic as compared to matric water potential. Soil Sci. Soc. Am. Proc. 35:465–469.
2. Ardakani, M. S., R. K. Schulz, and A. D. McLaren. 1974. A kinetic study of ammonium and nitrate oxidation in a soil field plot. Soil Sci. Soc. Am. Proc. 38:273–277.

3. Baver, L. D., W. H. Gardner, and W. R. Gardner. 1972. Soil physics. Fourth edition. John Wiley & Sons, Inc., New York.

4. Black, C. A., D. D. Evans, J. L. White, L. E. Ensminger, and F. E. Clark (ed.). 1965. Methods of soil analysis. Part 1, Agronomy No. 9. Am. Soc. of Agron., Madison, Wis.

5. Brooks, R. H., and A. T. Corey. 1966. Properties of porous media affecting fluid flow. J. Irrig. Drainage Div. Proc. Am. Soc. Civ. Eng. 92:62–88.

6. Brown, R. W., and B. P. VanHaveren. 1972. Psychrometry in water relations research. Utah Agric. Exp. Stn., Logan, Utah.

7. Campbell, G. S. 1977. An introduction to environmental biophysics. Springer-Verlag, New York.

8. ———, and M. D. Campbell. 1974. Evaluation of a thermocouple hygrometer for measuring leaf water potential in situ. Agron. J. 66:24–27.

9. ———, and W. H. Gardner. 1971. Psychrometric measurement of soil water potential: temperature and bulk density effects. Soil Sci. Soc. Am. Proc. 35:8–12.

10. ———, and G. A. Harris. 1977. Water relations and water use patterns for *Artemisia tridenta* Nutt. in wet and dry years. Ecology 58:652–659.

11. ———, and A. M. Wilson. 1972. Water potential measurements of soil samples. p. 142–149. *In* W. Brown, and B. P. VanHaveren (ed.) Psychrometry in water relations research. Utah Agric. Exp. Stn., Logan, Utah.

12. ———, W. D. Zollinger, and S. A. Taylor. 1966. Sample changer for thermocouple psychrometers: construction and some applications. Agron. J. 58:315–318.

13. Clothier, B. E., D. R. Scotter, and J. P. Kerr. 1977. Water retention in soil underlain by a coarse-textured layer: theory and a field application. Soil Sci. 123:392–399.

14. Cook, R. J., and R. I. Papendick. 1972. Influences of water potential of soils and plants on root disease. Ann. Rev. Phytopathol. 10:349–374.

15. ———, and ———. 1978. Role of water potential in microbial growth and development of plant disease, with special reference to postharvest pathology. Hort. Sci. 13(5):559–564.

16. ———, ———, and D. M. Griffin. 1972. Growth of two root-rot fungi as affected by osmotic and matric water potentials. Soil Sci. Soc. Am. Proc. 36:78–82.

17. Currie, J. A. 1965. Diffusion within soil microstructure a structural parameter for soils. J. Soil Sci. 16:279–289.

18. Gardner, W. R., F. N. Dalton, and R. F. Harris. 1972. Thermocouple psychrometry for the study of water relations of soil microorganisms. p. 150–153. *In* R. W. Brown and B. P. VanHaveren (ed.) Psychrometry in water relations research. Utah Agric. Exp. Stn., Logan, Utah.

19. ———, D. Hillel, and Y. Benyamini. 1970. Post irrigation movement of soil water: I. Redistribution. Water Resour. Res. 6:851–861; II. Simultaneous redistribution and evaporation. Water Resour. Res. 6:1148–1153.

20. Griffin, D. M. 1972. Ecology of soil fungi. Syracuse Univ. Press, New York.

21. ———, and G. Quail. 1968. Movement of bacteria in moist, particulate systems. Aust. J. Biol. Sci. 21:579–582.

22. Harris, R. F., W. R. Gardner, A. A. Adebeyo, and L. E. Sommers. 1970. Agar dish isopiestic equilibration method for controlling the water potential of solid substrates. Appl. Microbiol. 19:536–537.

23. Hillel, Daniel. 1971. Soil and water: physical principals and processes. Academic Press, New York.

24. Hsieh, J. J. C., W. H. Gardner, and G. S. Campbell. 1972. Experimental control of soil water content in the vicinity of root hairs. Soil Sci. Soc. Am. Proc. 36:418–421.

25. Lang, A. R. G. 1967. Osmotic coefficients and water potentials of sodium chloride solutions from 0 to 40 C. Aust. J. Chem. 20:2017–2023.

26. Letey, John, Jr., L. H. Stolzy, and W. D. Kemper. 1967. Soil aeration in irrigation of agricultural lands. p. 941–949. *In* R. M. Hagen, H. R. Haise, T. W. Edminster (ed.) Irrigation of agricultural lands. Agronomy 11, Academic Press, New York.

27. McLaren, A. D. 1970. Temporal and vectoral reactions of nitrogen in soil. Can. J. Soil Sci. 50:97–109.

28. Neumann, H. H., and G. W. Thurtell. 1972. A Peltier cooled thermocouple dewpoint hygrometer for in situ measurement of water potentials. p. 103–112. *In* R. W. Brown and B. P. VanHaveren (ed.) Psychrometry in water relations research. Utah Agric. Exp. Stn., Logan, Utah.

29. Olsen, S. R., and W. D. Kemper. 1968. Movement of nutrients to plant roots. Adv. Agron. 20:91–151.

30. Papendick, R. I., and G. S. Campbell. 1975. Water potential in the rhizosphere and plant and methods of measurement and experimental control. p. 34–49. *In* G. W. Bruehl (ed.) Biology and control of soil-born plant pathogens. Am. Phytopathol. Soc., St. Paul, Minn.

31. ————, V. L. Cochran, and W. M. Woody. 1971. Soil water potential and water content profiles with wheat under low spring and summer rainfall. Agron. J. 63:731–734.

32. Rawlins, S. L., and F. N. Dalton. 1967. Psychrometric measurement of soil water potential without precise temperature control. Soil Sci. Soc. Am. Proc. 31:297–301.

33. Richards, L. A. 1960. Advances in soil physics. 7th Int. Congr. Soil Sci. I:67–79.

34. Robinson, R. A., and R. H. Stokes. 1965. Electrolyte solutions. Butterworths, London.

35. Sabey, B. R. 1969. Influence of soil moisture tension on nitrate accumulation in soils. Soil Sci. Soc. Am. Proc. 33:263–266.

36. Scott, W. J. 1957. Water relations of food spoilage microorganisms. Adv. Food Res. 7:83–127.

37. Slatyer, R. O. 1967. Plant water relationships. Academic Press, London.

38. Stolzy, L. H., J. Letey, L. J. Klotz, and C. K. Labanauskas. 1965. Water and aeration as factors in root decay of *Citrus sinensis*. Phytopathology 55:270–275.

39. U.S. Salinity Laboratory Staff. 1954. Diagnosis and improvement of saline and alkali soils. USDA Agric. Handb. No. 60.

40. Wallace, H. R. 1958. Movement of eelworms. I. The influence of pore size and moisture content of the soil on the migration of larvae of the beet eelworm, *Heterodera schactii* Schmidt. Ann. Appl. Biol. 46:74–85.

41. Wiebe, H. H., G. S. Campbell, W. H. Gardner, S. L. Rawlins, J. W. Cary, and R. W. Brown. 1971. Measurement of plant and soil water status. Utah Agric. Exp. Stn. Bull. 484.

CHAPTER 2

Effect of Water Potential on Microbial Growth and Activity

R. F. HARRIS[1]

INTRODUCTION

Microbial tolerance to water stress varies widely, in general showing little relationship to classical taxonomy (Tables 1, 2). The mechanistic basis of microbial water relations has received increasing attention since Scott's (1957) classic water activity-based review, particularly in the areas of extreme xerotolerance (Brown, 1976, 1978; Rose, 1976; Dundas, 1977; Lanyi, 1978), food spoilage (Pitt, 1975; Leistner and Rodel, 1976), wood decay (Griffin, 1977), and the water relations of soil fungi and soil-borne pathogens (Griffin, 1969, 1978). In addition, a recent Dahlem workshop on "Life at low water activities" had a strong microbial orientation (Shilo, 1979). Current theories on mechanisms of water stress resistance by higher and lower plants, with particular emphasis on the role of turgor pressure, are reviewed by Hsiao et al. (1976).

Major recent advances in the biophysics and biochemistry of extreme xerotolerance have led to the introduction of the concept of compatible solutes, and of its extension as a general mechanism of intracellular water control to nonxerotolerant as well as xerotolerant microorganisms (Brown, 1976, 1978). Similarly, the chemiosmotic basis of energy coupling by *Halobacterium halobium* (Lanyi, 1978) has general implications regarding osmoregulation mechanisms used by non-photosynthetic as well as photosynthetic microorganisms under water stress.

Water activity (A_w) is still the most widely used parameter for characterizing microbial water relations (Brown, 1978), although the powerful mechanistic advantage of the additive nature of water potential (ψ) is gaining increasing recognition by microbiologists (Adebayo et al., 1971; Griffin, 1978).

This review evaluates the water potential relations of microorganisms in terms of compatible solute and energetic concepts of microbial growth under water stress. Particular emphasis is placed on physiological characterization of the different mechanistic strategies used

[1] Dep. of Soil Science and Bacteriology, Univ. of Wisconsin, Madison, WI 53706.

Table 1. Microbial tolernace to solute-controlled ψ stress.

Maximum tolerance					
		NaCl		Sucrose	
ψ	Aw	Molality	(W/V)	(W/W)	Microorganism†
bars		moles/ kg H_2O	——	% ——	
−15	0.99	0.33	2.0	17	Spiral bacteria: *Spirillum* (1,2) Gr⁻ bacterial rods: *Xanthomonas* (1); *Rhizobium* (3) Gr⁻ bacterial cocci: *Lamprodia* (1)
−40	0.97	0.87	5.2	33	Spirochete bacteria: *Treponema* (1) Spiral bacteria: *Spirillum* (1,2); *Campylobacter* (1) Gr⁻ rods: *Xanthomonas*, *Erwinia*, *Zoogloea* (1); *Pseudomonas* (2,4,5); *Achromobacter*, *Serratia* (2); S-oxidizers (7); nitrifiers (8); *Rhizobium* (3). Gr⁻ cocci: *Veillonella* (1) Gr⁺ rods: *Clostridium* (2,5); pathogenic *Bacillus* (1) Basidiomycete yeasts (2): *Hansenula* (4).
−70	0.95	1.48	8.8	44	Budding bacteria: *Hyphomicrobium* (1) Gr⁻ rods: *Vibrio* (1,2,5,6); *Escherichia* (5,6,9); *Salmonella* (5,6); *Flavobacterium*, *Klebsiella*, *Proteus*, *Shigella*, *Alcaligenes*, *Serratia* (5); *Pseudomonas* (5,10); *Citrobacter* (5,11). Gr⁺ rods: *Clostridium* (2,5); *Bacillus* (1,2,5); *Sporosarcina* (1). Actinomycetes: *Eubacterium*, *Bacteronema*, *Mycobacterium*, *Nocardia*, *Micromonospora* (1); *Streptomyces* (1,12,13). Basidiomycete yeasts and fungi (2). Phycomycete fungi: *Phytophthora*, *Phycomyces*, *Ophiobolus*, *Rhizoctonia* (13); *Mucor* (15). Algae: *Dunaliella tertiolecta* (2).
−100	0.93	2.05	12.3	52	Gr⁻ rods: *Flavobacterium* (1); *Vibrio* (5,16); *Serratia*, *Klebsiella* (6); *Aerobacter* (4); *Pseudomonas* (10, 16). *Escherichia*, *Salmonella* (16). Gr⁺ rods: *Clostridium* (4,6,16); *Bacillus* (2,6); *Arthrobacter* (9); *Lactobacillus* (2,4,5,6); *Microbacterium* (4,5). Actinomycetes: *Corynebacterium* (1); *Streptomyces* (1,12) Gr⁺ cocci: *Micrococcus* (1,6); *Aerococcus* (1); *Pediococcus* (1,2); *Streptococcus* (2,5,6). Yeasts: *Candida*, *Torula* (4) Phycomycete and imperfect fungi: *Rhizopus*, *Mucor* (4,5); *Botrytis* (4); *Fusarium* (11,13); *Cylindrocarpon*, *Cochliobolus* (13).
−150	0.90	2.92	17.3	60	Gr⁻ rods: *Salmonella* (4); *Vibrio* (1,2,5) Gr⁺ rods: *Bacillus* 4,6,16); *Lactobacillus* (5) Actinomycetes: *Corynebacterium* (5); *Streptomyces* (1,12) Gr⁺ cocci: *Micrococcus* (1,5,16); *Pediococcus* (1,5); *Streptococcus* (5); *Sarcina* (16); *Staphylococcus* (1). Yeasts: *Saccharomyces* (2,4,5); *Rhodotorula*, *Pichia*, *Hansenula* (5). Fungi: *Fusarium*, *Penicillium*, *Geastrum* (17)

(continued on next page)

Table 1. Continued.

Maximum tolerance					
		NaCl		Sucrose	
ψ	Aw	Molality	(W/V)	(W/W)	Microorganism†
bars		moles/ kg H_2O	—— %	——	
−200	0.86	3.70	21.7	65	Gr^- rods: *Vibrio* (4) Gr^+ cocci: *Micrococcus* (2,16); *Staphylococcus* (1,2,4, 5,6) Yeasts: *Endomyces* (2); *Saccharomyces* (2,4); *Candida, Debaromyces, Hanseniaspora, Torulopsis* (5) Ascomycete and imperfect fungi: *Cladosporium, Paecilomyces* (5); *Aspergillus* (13).
−250	0.83	4.41	25.5	70	Gr^+ cocci: *Micrococcus* (4) Ascomycete fungi: *Alternaria* (4); *Aspergillus* (4,5, 13,18,19); *Penicillium* (13,18,19); *Paecilomyces* (18,19). Yeasts: *Debaromyces* (2,18)
−350	0.76	5.68	--	75	Yeasts: *Saccharomyces* (2,5) Fungi: *Aspergillus, Penicillium* (4,5,13,18,19) Halophilic bacteria: *Halobacterium, Halococcus* (2,5) Halophilic actinomycetes: *Actinospora (2)* Halophilic algae: *Dunaliella* (2)
−500	0.69	NA	NA	80	Ascomycete fungi: *Aspergillus* (4,5,18,19); *Eurotium* (5,18); *Chrysosporium, Eremascus, Wallemia* (18,19).
−650	0.62	NA	NA	83	Ascomycete fungi: *Xeromyces* (4,18) Yeasts: *Saccharomyces* (5,18)

† References: 1. Buchanan and Gibbons (1975); 2. Brown (1976); 3. Steinborn and Roughley (1975); 4. Rose (1976); 5. Leistner and Rodel (1976); 6. Measures (1975); 7. Keller (1969); 8. Sindhu and Cornfield (1967); 9. Wasileski et al. (1978); 10. Prior (1978); 11. Harris (unpublished); 12. Wong and Griffin (1974); 13. Griffin (1972); 14. Kouyeas (1964); 15. Adebayo et al. (1971); 16. Marshall et al. (1971); 17. Wilson and Griffin (1975a); 18. Pitt (1975); 19. Pitt and Hocking (1977).

by nonxerotolerant and moderately tolerant microorganisms, and evaluation of the ecological implication of these strategies with respect to growth under solute- vs. matric-controlled ψ stress. The effect of ψ stress on the efficiency and rate determinants of microbial growth rate, survival, and activity under saturating and subsaturating substrate conditions, is reviewed, with emphasis on the energy-limiting conditions characteristic of soil environments. However, the emphasis of the review is on the principles controlling the effect of ψ on growth and activity of microorganisms, and the literature on microbial ψ relations is evaluated more in the context of principle illustration rather than from a comparative ψ tolerance standpoint. Specific examples of the role of water potential as a determinant of competitive growth survival and biochemical activity in soil environments are reviewed in other chapters (Griffin, Chapter 5; Cook and Duniway, Chapter 4; Sommers et al., Chapter 3).

Table 2. Microbial tolerance to matric-controlled ψ stress.

Maximum tolerance		Water film thickness†	Microbial activity in soil‡
ψ	Aw		
bars		μm	
−0.3	0.999	4	Movement of protozoa, zoospores, bacteria (1)
−1	0.999	1.5	
−5	0.996	0.5	
−15	0.99	30×10^{-4} (10 H_2O)	Nitrification, sulfur oxidation (1) Phycomycete fungal growth: *Achyla, Allomyces, Saprolegnia, Pythium* (2).
−40	0.97	$< 30 \times 10^{-4}$ ($<$ 10 H_2O)	Nitrification (3,4,5); Sulfur oxidation (6) *Bacillus* growth (7) Competitive bacterial and actinomycete growth (1) Phycomycete fungal growth: *Arthrobotrys, Pythium* (2); *Phytophthora* (2,8).
−100	0.93	$< 15 \times 10^{-4}$ ($<$ 5 H_2O)	Fungal growth: *Fusarium* (1,2); *Cylindrocarpon, Gliocladium, Absidia* (2); *Cochliobolus* (1); *Alternaria* (7)
−400	0.75	$< 9 \times 10^{-4}$ ($<$ 3 H_2O)	Fungal growth: *Rhizopus, Chaetomium, Aspergillus Scopulariopsis, Penicillium* (1,2).

† Numbers in parentheses are molecules of water.
‡ Reference: 1. Griffin (1972); 2. Kouyeas (1964); 3. Justice and Smith (1962); 4. Reichman et al. (1966); 5. Dubey (1968); 6. Moser and Olsen (1953); 7. Wilson and Griffin (1975b); 8. Adebayo and Harris (1971).

BIOPHYSICAL AND BIOCHEMICAL BASIS OF MICROBIAL WATER POTENTIAL RELATIONS

The primary response of microbial cells to water potential stress is dominated by the biophysical need of the cells to achieve thermodynamic water potential equilibrium with their environment (Adebayo et al., 1971; Brown, 1978; Griffin, 1978):

$$(\psi_t)_{sub} = (\psi_t)_{cells} \quad [1]$$

$$\Sigma(\psi)_{sub\ components} = \Sigma(\psi)_{cells\ components} \quad [2]$$

where $(\psi_t)_{sub}$ and $(\psi_t)_{cells}$ are the total water potential of the microbial substrate and microbial cells, respectively. The mechanisms by which microbial cells adjust the different components of their total ψ to maintain ψ equilibrium with their environment are generally considered to be largely biochemical in nature, reflecting varying degrees of osmoregulatory control over the forms and amounts of intracellular solutes comprising the solute ψ component of the cells (Brown, 1976, 1978). Accordingly, growth rate reduction responses to solute-controlled ψ stress are considered to be predominantly a reflection of inhibition of biochemical processes by excessively high intracellular solute concentrations, although increased osmoregulation energy costs under ψ stress may also be involved (Brown, 1976, 1978; Watson, 1970).

The solute transport properties of microbial cell membranes and walls play a major role in determining differences in the shock and growth response of microorganisms to ψ stress. The solute permeability of microbial membranes and walls varies with solute physical and chemical properties and cell membrane/wall physical structure and chemical composition. High molecular weight solutes such as proteins, polysaccharides, and long chain polyethylene glycol (PEG) are essentially impermeant with respect to both cell walls and cell membranes. Cell walls are relatively permeant to low M.W. solutes. However, diffusivity characteristics of low M.W. solutes with respect to biological membranes are highly variable. Nonionic solutes such as glycerol are generally more permeant than highly charged ionic solutes; K^+ tends to be much more permeant than Na^+ (Gradman, 1977; Lanyi, 1978). Osmoregulatory control of intracellular solute composition is based on restricted membrane permeability and the associated need for active carrier systems for solute transport. In contrast, biological membranes are highly permeant to water (Brown, 1978). Thus, the immediate shock response of microorganisms to ψ stress is dominated by water movement in response to the ψ gradient. For example, the immediate downshock response of a microorganism following transfer from a high ψ to a low ψ, eg −80 bar NaCl (1.78 *M*), medium will be to lose water, thereby increasing the intracellular solute concentration and resulting in a corresponding decrease in cell ψ; a secondary response, caused in part by cell dehydration-induced membrane distortion, is influx of NaCl in response to the NaCl concentration gradient across the membrane (high out, low in). The simplest mechanism of subsequent intracellular ψ adjustment to regain turgor and reestablish biochemically-functional concentrations of intermediary metabolites is to allow NaCl to passively equilibrate across the cell membrane so that the substrate −80 bar NaCl is counterbalanced by −80 bar NaCl in the cell. However, the energy-conserving advantages of passive equilibration under solute-controlled ψ stress are counterbalanced by potentially severe inhibitory effects of stress solutes such as NaCl on intracellular enzyme activity. An alternative physiological option involves exclusion of the ψ stress solutes and achievement of ψ adjustment by selective intracellular accumulation of solutes more compatible with cell enzymes (Brown, 1976, 1978). This option is inevitably energy-demanding since the cell is committed to establishment and maintenance of additional concentration gradients under ψ stress (Pirt, 1975; Thauer et al., 1977; Veldkamp, 1976). Specifically, for example, Na^+ efflux from cells under NaCl stress is accomplished at the expense of the proton motive force via a Na^+/H^+ antiport mechanism (Lanyi, 1978).

In summary, the basic response of a microorganism to ψ stress is dominated by the biophysical need for the organism to attain ψ equilibrium with its environment. However, prediction of the ultimate physiological response depends on an understanding of the nature as well as level of the ψ stress and on the mechanisms by which the cell adjusts the individual components of its intracellular water potential to attain total ψ equilibrium.

Specific Components of Microbial Substrate and Microbial Cell Water Potential

The different components of microbial substrate and microbial cell water potential are conveniently illustrated using a schematic of the ψ and solute balance of typical Gr^- and Gr^+ bacterial cells growing in minimal, high ψ liquid medium (Fig. 1).

COMPONENTS OF MICROBIAL SUBSTRATE WATER POTENTIAL

From a mechanistic standpoint, the total ψ of a microbial substrate is a function essentially of two basic components (Adebayo et al., 1971; Papendick and Campbell, Chapter 1):

$$(\psi_t)_{sub} = (\psi_s)_{sub} + (\psi_m)_{sub} \qquad [3]$$

where $(\psi_s)_{sub}$ is mechanistically defined from a biological cell standpoint as the substrate ψ component due to water sorption by dissolved solutes permeant (passive and/or active transport) with respect to cell wall/membranes, and $(\psi_m)_{sub}$ is the substrate ψ component due to water sorption by impermeant dissolved solutes, colloids and/or particulate surfaces.

Specific solute effects inevitably complicate interpretation of microbial response to ψ_s. In this regard, two broad categories of ψ_s are generally recognized: 1) Ionic solute-controlled ψ_s, such as NaCl-controlled ψ_s, with microbial ψ_s relations commonly defined in terms of molality or percent W/V of NaCl (Table 1; Brown, 1976, 1978) and 2) Nonionic solute controlled-ψ_s, such as sucrose-controlled ψ_s, with microbial response to ψ_s commonly defined in terms of percent W/W of sucrose (Table 1; Brown, 1976, 1978; Pitt, 1975). For the purpose of this review, the following subcomponents of $(\psi_s)_{sub}$ will be recognized:

$$(\psi_s)_{sub} = (\psi_{bs})_{sub} + (\psi_{ss})_{sub} \qquad [4]$$

where, for mechanistic convenience, $(\psi_{bs})_{sub}$ is defined as the ψ due to basal solutes in the growth medium, and (ψ_{ss}) is the ψ due to solutes which, because of their relatively high concentration, are defined as stress solutes.

Solute diffusion effects may complicate interpretation of microbial response to substrate ψ_m, depending on the mechanism of substrate ψ_m control. For impermeant solute-controlled ψ_m, e.g., polyethylene glycol(PEG)4000-controlled ψ_m (McAneney et al., 1980; Mexal and Reid, 1973), solute diffusion is relatively unimportant (assuming adequate mixing), and consequently use of impermeant solutes provides a useful means of characterizing the solute diffusion-independent, ψ_m relations of microorganisms (McAneney et al., 1980). Particulate matrix-controlled ψ_m (due to adsorptive or surface tension forces in solid substrates) is essentially negligible in aqueous substrates because of the inherently high water content of such substrates. For solid substrates, matrix-controlled ψ_m becomes increasingly more important as the substrate water content decreases. However, with decreasing substrate water content (θ_w), the cross-sectional

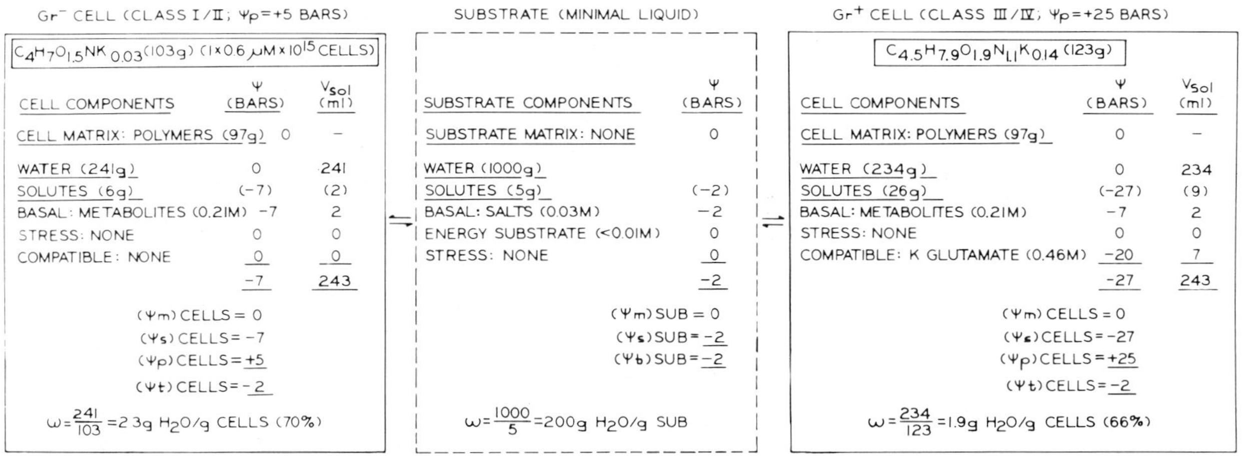

Gr^- CELL (CLASS I/II; Ψ_p=+5 BARS)

$C_4H_7O_{1.5}NK_{0.03}$ (103g) (1×0.6 μM×10^{15} CELLS)

CELL COMPONENTS	Ψ (BARS)	V_{sol} (ml)
CELL MATRIX: POLYMERS (97g)	0	–
WATER (241g)	0	241
SOLUTES (6g)	(−7)	(2)
BASAL: METABOLITES (0.21M)	−7	2
STRESS: NONE	0	0
COMPATIBLE: NONE	0	0
	−7	243

(Ψ_m) CELLS = 0
(Ψ_s) CELLS = −7
(Ψ_p) CELLS = +5
(Ψ_t) CELLS = −2

$\omega = \frac{241}{103} = 2.3$g H_2O/g CELLS (70%)

⇌

SUBSTRATE (MINIMAL LIQUID)

SUBSTRATE COMPONENTS	Ψ (BARS)
SUBSTRATE MATRIX: NONE	0
WATER (1000g)	
SOLUTES (5g)	(−2)
BASAL: SALTS (0.03M)	−2
ENERGY SUBSTRATE (<0.01M)	0
STRESS: NONE	0
	−2

(Ψ_m) SUB = 0
(Ψ_s) SUB = −2
(Ψ_b) SUB = −2

$\omega = \frac{1000}{5} = 200$g H_2O/g SUB

⇌

Gr^+ CELL (CLASS III/IV; Ψ_p=+25 BARS)

$C_{4.5}H_{7.9}O_{1.9}N_{1.1}K_{0.14}$ (123g)

CELL COMPONENTS	Ψ (BARS)	V_{sol} (ml)
CELL MATRIX: POLYMERS (97g)	0	–
WATER (234g)	0	234
SOLUTES (26g)	(−27)	(9)
BASAL: METABOLITES (0.21M)	−7	2
STRESS: NONE	0	0
COMPATIBLE: K GLUTAMATE (0.46M)	−20	7
	−27	243

(Ψ_m) CELLS = 0
(Ψ_s) CELLS = −27
(Ψ_p) CELLS = +25
(Ψ_t) CELLS = −2

$\omega = \frac{234}{123} = 1.9$g H_2O/g CELLS (66%)

Fig. 1. Schematic of the water potential (ψ), water content (ω), and intracellular and empirical composition of model Gr^- and Gr^+ cells in minimal liquid medium.

area of water available for microbial locomotion and solute diffusion is also decreased, thereby creating problems of cell mobility and restricted substrate supply to, and removal of metabolic by-products away from, the cell surface. In soil, matrix-controlled ψ_m reflects the size distribution of water-filled capillaries and thicknesses of water films over particulate surfaces. The effect of soil ψ_m on water distribution in soil is discussed by Papendick and Campbell (Chapter 1). In summary, the relationship between soil matrix-controlled ψ_m and soil water content is defined by the soil water retention isotherm properties of the soil, taking into account consideration of hysteresis effects. In terms of distribution of water in soil, soil ψ_m of -0.1 bar is normally associated with water saturation of soil capillaries $\leq 30\ \mu$ diam; -0.3 bar $< 4\ \mu$; and at $\psi_m < -5$ bars the soil water tends to be distributed as film only a few water molecules thick (Baver et al., 1972). The water content problem associated with decreasing matrix-controlled ψ_m is more severe for bacterial cells than for filamentous microorganisms (Table 2) since the latter are not dependent on water films for cell locomotion. However, an aqueous phase is presumably necessary at hyphal/substrate contact points to facilitate solute transport between the substrate and the cells; whether filamentous microorganisms capable of growth at low matric potentials (Table 2) overcome this problem by creating local zones of high solute concentrations (low ψ_s) at such contact points through excretion of low M.W. solutes and/or by extracellular enzyme action on the substrate is not known. Mechanistic interpretation of the effect of decreasing soil water on microbial growth and activity requires separation of the ψ per se, cell mobility and solute diffusion effects, and hence requires an understanding of the independent ψ per se, cell mobility and solute diffusion relations of the microbial population. The effects of soil water on bacterial cell mobility and solute diffusion are considered elsewhere (Griffin, Chapter 5; Papendick and Campbell, Chapter 1; Gardner and Harris, 1980). This review is concerned only with evaluation of the ψ per se relations of microorganisms.

A typical minimal, liquid microbial growth medium shows a basal salt-controlled $(\psi_{bs})_{sub}$ of -1 to -4 bars (Sommers et al., 1970; McAneney et al., 1980), largely determined by the phosphate buffer level of the medium: The medium in Fig. 1 reflects a basal salts composition (dominated by phosphate buffer) of about 0.03 *M* (Table A-1). Soil solutions in field-moist soils generally show > -1 bar $(\psi_{bs})_{sub}$. The ψ stress component, $(\psi_{ss})_{sub}$, is essentially negligible in commonly used liquid growth media and nonsaline soils. Solutes added to microbial substrates for characterization of solute ψ relations of microorganisms include electrolytes such as NaCl, KCl, and Na_2SO_4 (Tables A-2a, A-3a) and nonelectrolytes such as sucrose, mannitol, glucose, and glycerol (Tables A-2b, A-3b). Experimentally, molar-based solutions (Tables A-3a, A-3b) facilitate media preparation and subsampling on a volume basis, the simplest and most widely used approach for microbial culture mass balance analysis. The matrix-controlled $(\psi_m)_{sub}$ component is a function of adsorptive or surface tension forces in solid substrates and is essentially negligible in liquid media. Similarly, unless specifically incorporated into the medium, the impermeant solute-controlled ψ_m component is negligible in liquid media.

Table A-1. Empirical composition and molality vs. water potential (ψ) and water displacement (V) properties of some common microbial media components, intracellular ψ-controlling solutes, and chemically-related compounds.

Compound	Empirical composition‡	Molecular weight	Water potential for molalities (at 20 C)§ of:			Displacement volumes for molalities (at 20 C) of:		
			0.1 M	0.5 M	1.0 M	0.1 M	0.5 M	1.0 M
		g/mole	bars			ml/1,000 g H_2O		
			Mono-species					
Alcohols, polyols and sugar aldehydes								
Methanol	CH_4O (6e⁻)	32.0	−2.3	−12	−24	4	19	38
Ethanol	C_2H_6O (12e⁻)	46.1	−2.4	−12	−24	4	28	55
Ethylene glycol	$C_2H_6O_2$ (10e⁻)	62.1	−2.4	−12	−24	5	27	54
Propylene glycol	$C_3H_6O_2$ (14e⁻)	76.1	−2.5	−12	−25	7	35	70
Glycerol	$C_3H_8O_3$ (14e⁻)	92.1	−2.4	−12	−25	7	36	71
Arabitol	$C_5H_{12}O_5$ (22e⁻)	152.1	--	--	−25¶	--	--	104¶
Mannitol	$C_6H_{14}O_6$ (26e⁻)	182.2	−2.4	−12	−25	12	60	119
Glucose	$C_6H_{12}O_6$ (24e⁻)	180.2	−2.4	−12	−25	11	56	112
Sucrose	$C_{12}H_{22}O_{11}$ (48e⁻)	342.3	−2.5	−13	−26	21	105	212
Urea and amino acids								
Urea	CH_4ON_2	60.1	−2.4	−12	−23	4	23	44
λ amino butyrate	$C_4H_9O_2N$ (18e⁻)	103.1	--	--	−26††	--	--	65††
Proline	$C_5H_9O_2N$ (22e⁻)	115.1	−2.6‡‡	--	−26‡‡	--	--	65‡‡
			Di-species					
Inorganic salts								
Ammonium chloride	NH_4Cl	53.5	−4.5	−22	−44	4	18	37
Sodium chloride	NaCl	58.4	−4.5	−22	−44 (−46)	2	9	18
Potassium chloride	KCl	74.6	−4.5	−22	−43 (−44)	3	14	28
Sodium phosphate	NaH_2PO_4	120.0	−4.3	−20	−35 (−39)	3	16	35
Potassium phosphate	KH_2PO_4	136.1	−4.3	−20	-- (−36)	3	20	--
Magnesium sulfate	$MgSO_4$	120.4	−2.8	−13	−28 (−26)	0	2	8

(continued on next page)

Table A-1. Continued.

Compound	Empirical composition‡	Molecular weight	Water potential for molalities (at 20 C)§ of:			Displacement volumes for molalities (at 20 C) of:		
			0.1 M	0.5 M	1.0 M	0.1 M	0.5 M	1.0 M
		g/mole	bars			ml/1,000 g H_2O		
			Di-species					
Carboxylic and amino acid salts								
Sodium acetate	$NaC_2H_3O_2$ ($8e^-$)	82.0	−4.6	−23	−43	4	20	42
Potassium acetate	$KC_2H_3O_2$ ($8e^-$)	98.1	(−4.7)	(−24)	-- (−50)	--	--	--
Sodium butyrate	$NaC_4H_7O_2$ ($20e^-$)	110.1	(−4.7)	(−25)	-- (−53)	--	--	--
Potassium glutamate	$KC_5H_8O_4N$ ($18e^-$)	185.2	−4.4‡‡	--	−45‡‡	--	--	66‡‡
			Tri-species					
Calcium chloride	$CaCl_2$	111.0	−6.3	−36	−77 (−78)	1	11	23
Potassium phosphate	K_2HPO_4	174.2	−5.5	−26	-- (−50)	2	16	--
Sodium sulfate	Na_2SO_4	142.1	−5.8	−23	-- (−48)	1	9	--
Potassium oxalate	$K_2C_2O_4$ ($2e^-$)	166.2	−5.8	−27	−53	5	24	50
Sodium tartrate	$Na_2C_4H_4O_6$ ($10e^-$)	194.1	−5.7	−26	−48	6	31	65

† Data calculated from osmolalities given by Wolf et al. (1973). T = 293.16 K; D_w = 0.99823 g H_2O•ml H_2O^{-1}; R = 8.31432 joules•$mole^{-1}$•°K^{-1}.

‡ Numbers in parentheses are C-bound electron composition (Harris and Adams, 1979).

§ Numbers in parentheses are 25 C data calculated from osmotic coefficients given by Robinson and Stokes (1959). T = 289.16 K; D_w = 0.99707 g H_2O• ml H_2O^{-1}.

¶ Estimated based on mannitol and glycerol data: Analogous data for −20 bar arabitol are 0.81 M and V = 83.

†† Assumed to be similar to proline: Analogous data for −20 bar GAB are 0.81 M and V = 50.

‡‡ Measured experimentally by psychrometic (ψ) and volumetric (V) methods (Unpublished data): Analogous data for −20 bar proline and K^+ glutamate$^-$ are 0.81 M and 0.46 M and V = 50 and 30, respectively.

Table A-2a. Water potential vs. molality and water displacement volumes of some electrolyte solutions used commonly for characterizing microbial ψ relations.†

	Molality						Water displacement, 20 C		
	NaCl		Na_2SO_4		KCl				
ψ	25 C	20 C	25 C	20 C	25 C	20 C	NaCl	Na_2SO_4	KCl
bars	moles/1,000 g H_2O						ml/1,000 g H_2O		
−1	0.02	0.02	0.02	0.02	0.02	0.02	0	0	0
−5	0.11	0.11	0.09	0.09	0.11	0.11	2	1	3
−10	0.22	0.23	0.18	0.19	0.22	0.23	4	3	6
−15	0.33	0.34	0.28	0.30	0.34	0.34	6	5	9
−20	0.44	0.45	0.38	0.44	0.45	0.46	8	8	13
−25	0.55	0.57	0.49	--	0.56	0.58	10	--	16
−30	0.66	0.68	0.60	--	0.68	0.70	12	--	19
−40	0.87	0.90	0.82	--	0.90	0.94	16	--	26
−50	1.08	1.13	1.06	--	1.13	1.17	20	--	33
−60	1.28	1.35	1.29	--	1.35	1.42	25	--	41
−70	1.48	1.57	1.52	--	1.57	1.66	29	--	56
−80	1.68	1.78	1.74	--	1.78	1.90	33	--	--
−100	2.05	2.19	2.15	--	2.21	--	42	--	--
−125	2.50	2.69	2.62	--	2.72	--	52	--	--
−150	2.92	3.16	3.04	--	3.22	--	62	--	--
−175	3.32	3.61	3.42	--	3.71	--	72	--	--
−200	3.70	4.03	3.75	--	4.17	--	82	--	--
−250	4.41	4.81	--	--	--	--	99	--	--
−300	5.07	--	--	--	--	--	--	--	--
−350	5.68	--	--	--	--	--	--	--	--
−400	--	--	--	--	--	--	--	--	--

† Data for 25 C and 20 C calculated from Robinson and Stokes (1959) and Wolf et al. (1973), respectively. See also Table A-1.

In accordance with Eq. 2 and 3, the total ψ of the typical microbial growth medium in Fig. 1 is:

$$(\psi_t)_{sub} = (\psi_m)_{sub} + (\psi_{bs})_{sub} + (\psi_{ss})_{sub} \qquad [5.1]$$

$$(\psi_t)_{sub} = 0 + (-2 + 0) = -2 \text{ bars} \qquad [5.2]$$

COMPONENTS OF MICROBIAL CELL WATER POTENTIAL

The total water potential of microbial cells, $(\psi_t)_{cell}$, is a function of the dissolved solute, $(\psi_s)_{cell}$, the matric, $(\psi_m)_{cell}$, and the turgor pressure, $(\psi_p)_{cell}$, components:

$$(\psi_t)_{cell} = (\psi_m)_{cell} + (\psi_s)_{cell} + (\psi_p)_{cell} \qquad [6]$$

The cell ψ_m component reflects bound water (i.e. nonsolvent) but is considered to be of relatively minor importance in non-ψ-stressed growing cells (Adebayo et al., 1971; Griffin, 1978). The possible importance of this component in cells growing under ψ stress is unknown, but it presumably becomes progressively more important as ψ stress severity increases and may well play a substantial role under extreme ψ stress (Schobert, 1977).

Table A-2b. Water potential vs. molality and water displacement volumes of some nonelectrolyte solutions used commonly for characterizing microbial ψ relations.†

	Molality									Water displacement, 20 C			
	Sucrose			Mannitol (Sorbitol)		Glucose		Glycerol					
ψ	25 C		20 C	25 C	20 C	25 C	20 C	25 C	20 C	Sucr	Mann	Gluc	Glyc
bars	moles/1,000 g H_2O									ml/1,000 g H_2O			
−1	0.04	(0.04)	0.04	(0.04)	0.04	(0.04)	0.04	(0.04)	0.05	14	9	4	3
−5	0.20	(0.20)	0.20	(0.20)	0.21	(0.20)	0.21	(0.20)	0.21	42	24	23	15
−10	0.39	(0.39)	0.39	(0.40)	0.41	(0.40)	0.40	(0.40)	0.41	83	49	45	29
−15	0.58	(0.58)	0.58	(0.60)	0.61	(0.60)	0.61	(0.60)	0.61	122	73	68	43
−20	0.76	(0.76)	0.76	(0.80)	0.81	(0.80)	0.80	(0.80)	0.80	161	96	90	57
−25	0.94	(0.93)	0.93	(0.99)	--	(0.99)	1.00	(1.00)	1.00	198	--	111	71
−30	1.10	(1.10)	1.10	(1.18)	--	(1.19)	1.19	(1.21)	1.19	232	--	133	85
−40	1.43	(1.43)	1.42	(1.56)	--	(1.57)	1.56	(1.60)	1.57	302	--	175	111
−50	1.74	(1.74)	1.72	(1.93)	--	(1.95)	1.91	(2.00)	1.94	365	--	215	138
−60	2.03	(2.04)	1.99	(2.30)	--	(2.33)	--	(2.40)	2.30	425	--	--	164
−70	2.31	(2.34)	--	(2.67)	--	(2.70)	--	(2.79)	2.66	--	--	--	189
−80	2.59	(2.62)	--	(3.01)	--	(3.07)	--	(3.18)	3.01	--	--	--	214
−100	3.11	(3.17)	--	(3.73)	--	(3.80)	--	(3.96)	3.71	--	--	--	264
−125	3.73	(3.81)	--	(4.60)	--	(4.70)	--	(4.94)	4.57	--	--	--	325
−150	4.33	(4.42)	--	(5.45)	--	(5.58)	--	(5.91)	5.42	--	--	--	386
−175	4.90	(5.01)	--	(6.28)	--	(6.46)	--	(6.89)	6.29	--	--	--	447
−200	5.47	(5.58)	--	(7.11)	--	(7.32)	--	(7.85)	7.14	--	--	--	509
−250	--	(6.67)	--	(8.73)	--	(9.03)	--	(9.80)	--	--	--	--	--
−300	--	(7.70)	--	(10.32)	--	(10.71)	--	(11.75)	--	--	--	--	--
−350	--	(8.09)	--	(11.89)	--	(12.39)	--	(13.71)	--	--	--	--	--
−400	--	(9.65)	--	(13.45)	--	(14.06)	--	(15.70)	--	--	--	--	--
−450	--	(10.58)	--	(15.01)	--	(15.73)	--	(17.71)	--	--	--	--	--
−500	--	(11.50)	--	(16.56)	--	(17.40)	--	(19.74)	--	--	--	--	--
−550	--	(12.39)	--	(18.11)	--	(19.08)	--	(21.80)	--	--	--	--	--
−600	--	(13.27)	--	(19.67)	--	(20.77)	--	(23.90)	--	--	--	--	--
−650	--	(14.14)	--	(21.23)	--	(22.47)	--	(26.03)	--	--	--	--	--

† Data for 25 C and 20 C were calculated from Robinson and Stokes (1959) and Wolf et al. (1973), respectively. Numbers in parentheses were calculated from the equation of Norrish (1966). See also Table A-1.

Table A-3a. Water potential vs. water activity and molal- and molar-based dry weights of some electrolyte solutions used commonly for characterizing microbial ψ relations.†

Water stress			Molal-based solute weight						20 C Molar-based (per liter solution)					
	Aw		NaCl		Na_2SO_4		KCl		NaCl		Na_2SO_4		KCl	
ψ	25 C	20 C	25 C	20 C	25 C	20 C	25 C	20 C	Solute weight	H_2O volume	Solute weight	H_2O volume	Solute weight	H_2O volume
bars			g/1,000 g H_2O						g/l	ml/l	g/l	ml/l	g/l	ml/l
−1	0.9993	0.9993	1.3	1.2	2.4	2.3	1.6	1.6	1.2	1,000	2.3	1,000	1.6	999
−5	0.9964	0.9963	6.4	6.5	12.1	12.3	8.2	8.3	6.4	998	12.3	999	8.3	997
−10	0.9927	0.9926	12.8	13.2	25.3	26.6	16.6	16.9	13.1	996	26.5	997	16.7	994
−15	0.9891	0.9890	19.3	19.7	39.4	42.6	25.0	25.6	19.6	994	42.3	995	25.3	991
−20	0.9855	0.9853	25.7	26.4	54.1	62.0	33.5	34.3	26.1	992	61.4	992	33.8	987
−25	0.9819	0.9817	32.1	33.0	69.3	--	42.0	43.1	32.7	990	--	--	42.4	984
−30	0.9784	0.9780	38.4	39.7	84.7	--	50.4	52.0	39.1	988	--	--	50.9	981
−40	0.9713	0.9708	50.8	52.8	116.8	--	67.2	69.8	51.9	984	--	--	67.9	974
−50	0.9642	0.9637	63.1	65.8	150.0	--	84.0	87.6	64.4	980	--	--	84.6	968
−60	0.9572	0.9565	74.9	78.8	183.0	--	100.5	105.9	76.2	976	--	--	100.7	961
−70	0.9503	0.9495	86.4	91.4	215.1	--	116.7	123.8	88.3	972	--	--	117.4	954
−80	0.9434	0.9425	97.9	103.9	247.1	--	132.9	141.7	100.4	968	--	--	134.0	947
−100	0.9297	0.9286	119.8	128.1	306.0	--	164.5	--	122.8	960	--	--	--	--
−125	0.9129	0.9116	145.9	157.1	372.8	--	202.8	--	149.0	950	--	--	--	--
−150	0.8964	0.8949	170.5	184.6	432.4	--	240.0	--	173.4	941	--	--	--	--
−175	0.8802	0.8785	193.9	210.7	485.3	--	276.6	--	196.2	933	--	--	--	--
−200	0.8644	0.8624	216.1	235.5	532.6	--	311.0	--	217.3	925	--	--	--	--
−250	0.8334	0.8310	257.7	281.3	--	--	--	--	255.4	910	--	--	--	--
−300	0.8036	0.8008	296.2	--	--	--	--	--	--	--	--	--	--	--
−350	0.7748	0.7717	332.1	--	--	--	--	--	--	--	--	--	--	--
−400	0.7471	0.7437	--	--	--	--	--	--	--	--	--	--	--	--

† Data for 25 C and 20 C were calculated from Robinson and Stokes (1959) and Wolf et al. (1973), respectively. See also Table A-1.

Table A-3b. Water potential vs. water activity and molal- and molar-based dry weights of some nonelectrolyte solutions used commonly for characterizing microbial ψ relations.†

Water stress		Molal-based solute weight								25 C Solute weight %				20 C Molar-based (per liter solution)							
		Sucrose		Mannitol (Sorbitol)		Glucose		Glycerol						Sucrose		Mannitol		Glucose		Glycerol	
ψ	Aw 25 C	25 C	20 C	25 C	20 C	25 C	20 C	25 C	20 C	Sucr	Mann	Gluc	Glyc	Sol wt	H_2O vol	Sol wt	H_2O vol	Sol wt	H_2O vol	Sol wt	H_2O vol
bars		g/1,000 g H_2O								g/100 g solution				g/l	ml/l	g/l	ml/l	g/l	ml/l	g/l	ml/l
−1	0.9993	13.8 (13.7)	14.0	(7.3)	7.5	(7.2)	7.4	(3.7)	4.3	(1.4)	(0.7)	(0.7)	(0.4)	13.8	986	7.4	991	7.3	995	3.8	998
−5	0.9964	68.1 (68.5)	68.6	(36.4)	37.5	(36.0)	37.0	(18.4)	19.1	(6.4)	(3.5)	(3.5)	(1.8)	65.7	959	36.6	976	36.1	978	18.8	986
−10	0.9927	134.2 (133)	135.1	(72.9)	74.3	(72.1)	72.8	(36.8)	37.6	(11.8)	(6.8)	(6.7)	(3.6)	124.5	923	70.8	954	69.5	957	36.5	972
−15	0.9891	198.2 (199)	198.5	(109)	111.1	(108)	109.3	(55.3)	55.8	(16.6)	(9.9)	(9.8)	(5.2)	176.5	891	103.4	932	102.2	936	53.4	959
−20	0.9855	260.1 (260)	260.1	(146)	147.3	(144)	144.2	(73.7)	73.8	(20.6)	(12.7)	(12.6)	(6.9)	223.1	862	134.1	912	132.1	918	69.7	946
−25	0.9819	319.9 (318)	319.7	(180)	--	(178)	179.3	(92.1)	91.6	(24.2)	(15.3)	(15.1)	(8.4)	266.5	835	--	--	161.0	890	85.4	934
−30	0.9784	377.9 (377)	377.1	(215)	--	(214)	214.1	(111)	109.7	(27.4)	(17.7)	(17.7)	(10.0)	305.2	811	--	--	188.6	882	100.9	922
−40	0.9713	489.2 (489)	486.3	(284)	--	(283)	280.7	(147)	144.3	(32.9)	(22.1)	(22.1)	(12.8)	373.1	769	--	--	238.5	851	129.6	900
−50	0.9642	595.1 (596)	587.6	(352)	--	(351)	344.7	(184)	178.3	(37.3)	(26.0)	(26.0)	(15.6)	429.8	733	--	--	283.2	823	156.5	879
−60	0.9572	696.3 (698)	682.4	(419)	--	(420)	--	(221)	212.0	(41.1)	(29.5)	(29.6)	(18.1)	478.4	702	--	--	--	--	181.9	859
−70	0.9503	791.4 (801)	--	(486)	--	(487)	--	(257)	277.4	(44.5)	(32.7)	(32.7)	(20.4)	--	--	--	--	--	--	228.1	824
−80	0.9434	886.4 (897)	--	(548)	--	(553)	--	(293)	309.5	(47.3)	(35.4)	(35.6)	(22.7)	--	--	--	--	--	--	269.8	791
−100	0.9297	1,066 (1,085)	--	(680)	--	(685)	--	(365)	341.5	(52.0)	(40.5)	(40.6)	(26.7)	--	--	--	--	--	--	317.0	754
−125	0.9129	1,278 (1,304)	--	(838)	--	(847)	--	(455)	420.7	(56.6)	(45.6)	(45.9)	(31.3)	--	--	--	--	--	--	359.8	722
−150	0.8964	1,482 (1,513)	--	(993)	--	(1,006)	--	(544)	499.3	(60.2)	(49.8)	(50.1)	(35.3)	--	--	--	--	--	--	399.4	691
−175	0.8802	1,679 (1,715)	--	(1,144)	--	(1,164)	--	(635)	578.2	(63.2)	(53.4)	(53.8)	(38.8)	--	--	--	--	--	--	435.5	663
−200	0.8644	1,871 (1,910)	--	(1,295)	--	(1,319)	--	(723)	657.8	(65.6)	(56.4)	(56.9)	(42.0)	--	--	--	--	--	--	--	--
−250	0.8334	-- (2,283)	--	(1,591)	--	(1,627)	--	(903)	--	(69.5)	(61.4)	(61.9)	(47.4)	--	--	--	--	--	--	--	--
−300	0.8036	-- (2,636)	--	(1,880)	--	(1,930)	--	(1,082)	--	(72.5)	(65.3)	(65.9)	(52.0)	--	--	--	--	--	--	--	--
−350	0.7748	-- (2,975)	--	(2,166)	--	(2,233)	--	(1,263)	--	(74.8)	(68.4)	(69.1)	(55.8)	--	--	--	--	--	--	--	--
−400	0.7471	-- (3,303)	--	(2,451)	--	(2,534)	--	(1,446)	--	(76.8)	(71.0)	(71.7)	(59.1)	--	--	--	--	--	--	--	--
−450	0.7204	-- (3,622)	--	(2,735)	--	(2,835)	--	(1,631)	--	(78.4)	(73.2)	(73.9)	(62.0)	--	--	--	--	--	--	--	--
−500	0.6946	-- (3,936)	--	(3,017)	--	(3,135)	--	(1,818)	--	(79.7)	(75.1)	(75.8)	(64.5)	--	--	--	--	--	--	--	--
−550	0.6697	-- (4,241)	--	(3,300)	--	(3,438)	--	(2,008)	--	(80.9)	(76.7)	(77.5)	(66.8)	--	--	--	--	--	--	--	--
−600	0.6458	-- (4,542)	--	(3,584)	--	(3,743)	--	(2,201)	--	(82.0)	(78.2)	(78.9)	(68.8)	--	--	--	--	--	--	--	--
−650	0.6227	-- (4,840)	--	(3,868)	--	(4,049)	--	(2,397)	--	(82.9)	(79.5)	(80.2)	(70.6)	--	--	--	--	--	--	--	--

† Data for 25 C and 20 C were calculated from Robinson and Stokes (1959) and Wolf et al. (1973), respectively; see also Table A-1. Numbers in parentheses were calculated from the equation of Norrish (1966), using K_2 values of 2.60 for sucrose, 0.85 for sorbitol and assuming 0.85 for mannitol, 0.70 for glucose and fructose, and 0.38 for glycerol.

Establishment of matric potentials in cell walls, especially under even mild ψ_m stress, is not unreasonable (McAneney et al., 1980) and will be discussed later. The $(\psi_s)_{cells}$ component is, as for $(\psi_s)_{sub}$, a function of the sum of different subcomponents:

$$(\psi_s)_{cells} = (\psi_{bs})_{cell} + (\psi_{ss})_{cell} + (\psi_{is})_{cell} + (\psi_{cs})_{cell} \quad [7]$$

where $(\psi_{bs})_{cell}$ is, for mechanistic convenience, defined as the ψ component due to basal intermediary metabolites; $(\psi_{ss})_{cell}$ is the ψ due to intracellular accumulation of stress solutes present in high (stress) concentration in the medium; $(\psi_{is})_{cell}$ is the ψ due to ψ stress-induced, intracellular accumulation of compatible solutes; and $(\psi_{cs})_{sub}$ is the ψ component due to constitutive intracellular accumulation of compatible solutes.

The basal intermediary metabolites comprising $(\psi_{bs})_{cell}$ are a diverse mixture of amino, nucleic, and carboxylic acids, carbohydrates, etc.; the net ionic charge of the cytoplasm is −ve, and K^+ is the dominant counterbalancing cation. The basal metabolites of Gr^- cells comprise about 6% of the cell dry weight (Mitchell and Moyle, 1956). For a typical cell water content of 2.32 g H_2O/g cells (about 70%), the cytoplasmic concentration of basal metabolites is thus 25.9 g/1,000 g H_2O; assuming an average F.W. of 125 g/mole, this converts into a molal concentration of about 0.2 *M* basal metabolites existing largely in the form of K^+ salts of amino and carboxylic acids and lesser amounts of uncharged organic metabolites. The resultant $(\psi_{bs})_{cells}$ for such a mixture is about −7 bars (Table A-1; Fig. 1; Mitchell and Moyle, 1956). A similar $(\psi_{bs})_{cells}$ component is assumed for Gr^+ cells (Fig. 1).

For non ψ-stressed cells the inducible compatible solute component is zero (Fig. 1). Under ψ stress, microbial cells may selectively accumulate compatible solutes such as amino acids (procaryotes and some photosynthetic eucaryotes) and polyols (eucaryotes) as a mechanism of achieving ψ equilibrium with their environment (Epstein and Schultz, 1968; Brown, 1976, 1978; Hsiao et al., 1976; Schobert, 1977). Since microbial production of such solutes only occurs in response to a ψ stress, the term inducible compatible solute would appear appropriate. Experimental problems (discussed in more detail later) of contamination-free cell separation and water content determination of microorganisms grown under ψ stress severely complicate calculation of intracellular solute concentrations (Brown, 1976, 1978; Griffin, 1978) and thus the following compatible solute boundary levels should be viewed only as best estimates. Bacteria accumulate K^+ glutamate$^-$ as a primary inducible compatible solute up to an intracellular level of 0.4 to 0.5 *M* (about −20 bars, Table A-1), followed by further accumulation of about 1 *M* (about −20 bars, Table A-1) of neutral amino acids such as proline or gamma amino butyrate (Measures, 1975; Brown, 1976; Koujima et al., 1978). Polyols such as glycerol, the favored inducible compatible solutes of eucaryotic microorganisms, accumulate up to levels of about −60 bars (2.3 *M*, Table A-2b) under ψ stress (Brown, 1978). Accordingly, under ψ stress conditions where the substrate ψ exceeds the inducible compatible solute limit, other mechanisms of intracellular ψ control must also be used (e.g. uptake of substrate stress solutes such as NaCl, if available). The biochemical basis of compatible solute action has been investigated extensively and is beyond the scope of this review: In summary, compatible solutes show minimal enzyme inhibition, a phenomenon related at least in part to enzyme stabilization by interactions between enzyme and compatible solute hydrophobic groups (Schobert, 1977; Brown, 1976, 1978) and, for K^+ replacement of Na^+ under NaClψ stress, to the decreased disruptive effect on water structure of K^+ (as a compatible cation) vs. Na^+ (Lanyi, 1978).

The role of constitutive compatible solutes is less clear-cut than that of inducible compatible solutes, but as discussed later probably involves turgor pressure as well as classical compatible solute action considerations. Gr^+ bacteria and certain thick-walled eucaryotic microorganisms characteristically show high levels of compatible solutes that may be considered constitutive in nature since they are present independent of ψ stress (Fig. 1). Constitutive accumulation of metabolites such as K^+ glutamate$^-$ by Gr^+ bacteria has long been known (Mitchell and Moyle, 1956; Christian and Waltho, 1964), with levels commonly falling in the 0.4 to 0.5 *M* (about -20 bars, Table A-1) range (Measures, 1975; Brown, 1976; Makemson and Hastings, 1979). Similarly, certain eucaryotic yeasts accumulate comparable ψ levels of arabitol (-20 bars $\simeq$ 0.8 *M*, Table A-1) (Brown, 1978) and/or glutamate (Tempest et al., 1970) as constitutive compatible solutes.

The turgor pressure component, $(\psi_p)_{cell}$, of microbial cell ψ is the internal hydrostatic pressure developed in cells due to a difference between the media ψ and the intracellular solute plus matric ψ (Eq. 6). For the unstressed Gr^- cells growing in basal medium of $(\psi_t)_{sub} = -2$ bars (Fig. 1, Eq. 6):

$$(\psi_p)_{cells} = (\psi_t)_{cells} - (\psi_s)_{cells} - (\psi_m)_{cells} \qquad [8.1]$$
$$= -2 - (-7) - 0 = +5 \text{ bars}$$

This relatively low turgor pressure is typical of that shown by Gr^- cells (Mitchell and Moyle, 1956) and many fungi (Adebayo and Harris, 1971; Griffin, 1978). Whether turgor pressure is under active physiological control or whether it is merely an indirect reflection of physiological needs for maintaining relatively high intermediary metabolite concentrations to drive biochemical reactions is not clear (Brown, 1976; Hsiao et al., 1976). However, even if maintenance of an optimum turgor pressure is not under active control, it is unreasonable to expect growth to proceed in the absence of at least a slightly positive ψ_p.

A relatively very high turgor pressure results from the accumulation of -20 bar K^+ glutamate$^-$ as a constitutive compatible solute by nonstressed Gr^+ cells growing in basal medium of $(\psi_t)_{sub} = -2$ bars (Fig. 1, Eq. 8.1):

$$(\psi_p)_{cells} = -2 - (-7) - (-20) + 0 = +25 \text{ bars} \qquad [8.2]$$

The high turgor pressure of Gr^+ cells is well established (Mitchell and Moyle, 1956; Christian and Waltho, 1964). Such high turgor pressures explain the need for Gr^+ cells and similar constitutive compatible solute-accumulating microorganisms to possess strong, thick cell walls. The functional rationale for this phenomenon has received little attention but certain implications are evident. In environments subjected to fluctuating ψ, such as soil, a high ψ_p buffers the cell against transient downshock ψ stress. From a growth standpoint, a high ψ_p may reflect a need for high internal hydrostatic pressure for (thick-walled) cell division. Alternatively, the high ψ_p in high ψ media might reflect atypical growth conditions: if the natural habitat is characterized by a low ψ, a correspondingly lower ψ_p would prevail in practice. This may explain the existence of an optimum substrate $\psi < -5$ bars shown by certain organisms (Scott, 1957; Sommers et al., 1970; Griffin, 1978). Similarly, the failure of certain xerotolerant fungi to grow in high ψ media (Scott, 1957), may be related to the generation of excessively high ψ_p due to the presence of high levels of constitutive compatible solutes [e.g. for a $(\psi_s)_{cells}$ of -47 bars associated with the presence of -40 bar constitutive compatible solute, the ψ_p would increase from $+5$ bars for a $(\psi_s)_{sub} = -42$ bars, to $+45$ bars for a $(\psi_s)_{sub} = -2$ bars]. Specific effects of different mechanistic options of ψ_p modulation on transient and growth response to ψ stress are evaluated later.

COMPONENTS OF MICROBIAL CELL WALL WATER POTENTIAL

The primary role of cell walls is to protect the cell, and in particular to provide structural rigidity necessary for prevention of plasmoptys due to turgor developed within the protoplast. From a ψ standpoint, the cell wall is in equilibrium with its surrounding substrate environment and its enclosed protoplast:

$$(\psi_t)_{sub} = (\psi_t)_{wall} = (\psi_t)_{protoplast} \qquad [9]$$

where $(\psi_t)_{wall}$ and $(\psi_t)_{protoplast}$ are the total ψ of the cell wall and protoplast, respectively.

Cell walls generally show a porous structure with limited (compared to the membrane-bound protoplasts) selective permeability due to the pore size distribution and charge characteristics of the cell wall polymers. The total ψ of cell walls is a function of a solute $[(\psi_s)_{wall}]$, a matric $[(\psi_m)_{wall}]$ and a pressure $[(\psi_p)_{wall}]$ component:

$$(\psi_t)_{wall} = (\psi_s)_{wall} + (\psi_m)_{wall} + (\psi_p)_{wall} \qquad [10]$$

The $(\psi_s)_{wall}$ and $(\psi_m)_{wall}$ components are comparable mechanistically to the corresponding whole cell ψ components. The $(\psi_p)_{wall}$ component is analogous to the swelling pressure in an inert colloidal system such as a clay suspension (McAneney et al., 1980). Current knowledge is inadequate to allow precise resolution of the relative importance of the different ψ_{wall} components under ψ stress. However, under non-ψ-stressed and ψ_s stressed conditions, cell wall pores are presumably filled with an aqueous solution comparable in composition to the surrounding medium.

Under such water-saturated conditions, $(\psi_m)_{wall}$ is probably insignificant, at least for low and moderate solute ψ stress. However, as solute ψ stress increases, the possible contribution of $(\psi_m)_{wall}$ should not be ignored (Schobert, 1977).

The cell wall ψ_m and ψ_p components become highly important for cells under substrate ψ_m stress because of the inability of the cell wall (unlike the microbial protoplast) to maintain hydration under ψ_m stress by selective accumulation of solutes. Thus, under ψ_m stress, the corresponding decrease in $(\psi_t)_{wall}$ must be accommodated by cell wall dehydration as cell wall pores empty according to the sorption moisture isotherm characteristics of the walls. For a rigid cell wall structure, the water-filled pore size distribution at low ψ_m will be confined to microcapillaries, and solute transport may be severely inhibited because of water channel discontinuities. For a pliable cell wall structure, dehydration-induced cell wall contraction will result in a downshift in total pore size distribution, again potentially inhibiting solute transport, although the water channel discontinuity problem may not be as severe. Differences between Gr^- and Gr^+ cell wall matric ψ relations are evaluated by McAneney et al. (1980).

The above discussion assumes no direct interaction between the ψ_s components of the protoplast and the ψ_s components of the cell wall. This is not necessarily a valid assumption, although direct information to the contrary is not available. However, especially for cells accumulating compatible solutes, it is not inconceivable that at least a certain amount of compatible solute efflux could occur to allow cell wall ψ equilibration as a function of dissolved solutes rather than matric mechanisms. Such a mechanism would result in massive leakage of compatible solutes into the liquid growth medium under impermeant solute-controlled ψ stress, but this loss would be attenuated considerably in substrates of matrix-controlled ψ stress because of the relatively low water contents and static (from a mixing standpoint) condition characteristic of such systems.

CHARACTERIZATION OF MICROBIAL WATER RELATIONS

Effect of Water Potential on Cell Composition

Characterization of microbial strategies for growth under ψ stress is facilitated by the fact that cell composition is dependent on the osmoregulatory mechanism used for intracellular ψ control. In addition, an understanding of the effect of ψ on cell composition is essential for mechanistic evaluation of microbial tolerance to ψ shock and energy cost interpretation of the effect of ψ stress on growth yield. However, experimental attainment of valid chemical composition, dry weight, and water content (necessary for intracellular solute concentration calculation) data for cells growing under ψ stress is complicated seriously by the severe technical problem of separating the cells from their growth medium in such a way as to remove growth medium contaminants without changing the composition of the cells (Brown, 1976, 1978; Griffin, 1978). The common procedure of water-washing cells prior to composition analysis creates serious problems because of inevitable ψ upshock-induced, intracellular solute loss. Cell ψ adjustment to wash water ($\psi \simeq 0$) is rapid: Water moves into the cells, which either burst or leak out intracellular solutes, particularly ψ stress-related inducible compatible and stress solutes, until ψ equilibrium is attained. Partial washing results in ill-defined plasmoptys and cell leakage. Washing with isotonic solutions (Watson, 1970; Brown, 1978) prevents ψ upshock but does not guarantee nonoccurrence of cell leakage, especially if washing is prolonged and the ψ-controlling solutes of the wash solution are not the same as the ψ-controlling components of the substrate. Cell contamination problems related to wash solution carry-over become increasingly more severe as substrate ψ (and thus isotonic wash solution ψ) increases, and this seriously hampers, if not prevents entirely, attainment of valid dry weight data. For aerial mycelia-producing fungi, direct harvesting, analogous to the harvesting of the aerial portions of higher plants, allows contamination-free cell separation from agar substrates (Adebayo et al., 1971).

Due to the limited scope of currently available information on ψ stress-cell composition interrelations (Brown, 1976, 1978; Griffin, 1978), the effect of different ψ stress responses on cell water content and empirical composition is analyzed for background reference in Appendix B. Resulting data are summarized in Tables A-4 and A-5 and Fig. 2.

Osmoregulation-Based Classification of Microbial Water Relations

Microorganisms may be classified from a water relation standpoint according to their physiological potential for osmoregulation via compatible solute production (Table 3; Fig. 2; Table A-5). Organisms incapable of either inducible or constitutive solute production (Class I) are potentially the most xerosensitive, particularly with respect to intolerance to ψ perturbations; at the other extreme of the physiological spectrum are organisms (Class IV) capable of withstanding extreme and highly perturbed ψ stress by virtue of their ability to produce high levels of both inducible and constitutive compatible solutes. Implicit in this classification approach is that the strategies used by microorganisms for growth and survival under ψ stress reflect the nature and extent of ψ perturbations of their natural habitats and, from a predictive standpoint, define the potential of microorganisms for growth and survival as a function of the specific ψ perturbation and solute properties of foreign habitats.

The following section evaluates the shock and growth response of each physiological class to solute ψ as compared to matric ψ stress and reviews evidence supporting assignment of representative microorganisms to specific classes. Examples of characteristic transient and growth-related cell composition changes in response to -80 bar NaCl stress are illustrated in Fig. 2 for typical procaryotic representatives of the four different classes, and comparable data for a broader representation of organisms growing under -40 and -80 bar NaCl stress are summarized for reference in Table A-5. It should be emphasized that the "maximum" levels of specific inducible and compatible solutes identified in Tables 3 and A-4 and Fig. 2 are highly tentative and should be considered only as best estimates.

Table A-4. Estimation of the effect of ψ stress on the intracellular solute, water content, and empirical composition of a microbial cell.

Cell component	ψ†	Intracellular solution			Relative concentration (moles) and empirical formula of cell components	Relative contribution of cell components to cell biomass‡		
		M	Displacement volume			Dry weight	C-bound electrons	Cell N
	bars	moles/ Kg H_2O	ml/ Kg H_2O	ml/mole cells		g/mole cells	e^-eq/mole cells	moles N/ mole cells
					Unstressed Gr^- cell			
Particulate phase	0				0.95 ($C_4H_7O_{1.5}N$ + 9.3 g)	97.3 (102.4)	16.15 (17.0)	0.95 (1.0)
Basal solutes	−7	0.21	10	2.4	0.05 ($C_4H_7O_{1.5}NK_{0.6}$ + 8.1 g)	6.2 (124)	0.85 (17.0)	0.05 (1.0)
Water	0		1,002	240.6		[240]		
Total†	−7		1,012	243	1.00 ($C_4H_7O_{1.5}NK_{0.3}$ + 9.3 g)	103.5 (103.5)	17.00 (17.0)	1.00 (1.0)
		Gr^- cell under −80 bar NaCl stress: K^+ glutamate$^-$ and proline induced as compatible solutes						
Particulate phase	0				0.95 ($C_4H_7O_{1.5}N$ + 9.3 g)	97.3 (102.4)	16.15 (17.0)	0.95 (1.0)
Basal solutes	−7	0.21	10	2.2	0.046 ($C_4H_7O_{1.5}NK_{0.6}$ + 8.1 g)	5.7 (124)	0.78 (17.0)	0.046 (1.0)
K^+ glutamate$^-$	−20	0.46	30	6.6	0.101 ($C_5H_8O_4NK$)	18.7 (185.2)	1.82 (18.0)	0.101 (1.0)
Proline	−20	0.81	50	11.0	0.177 ($C_5H_9O_2N$)	20.4 (115.1)	3.89 (22)	0.177 (1.0)
NaCl	−40	0.90	16	3.5	0.198 (NaCl)	11.6 (58.4)	0 (0)	0 (0)
Water	0		1,002	219.8		[219]		
Total cell†	−87		1,108	243	$C_{5.38}H_{9.37}O_{2.24}N_{1.28}K_{0.13}(NaCl)_{0.20}$	153.7 (153.7)	22.64 (22.64)	1.28 (1.28)
					$C_{4.17}H_{7.32}O_{1.75}NK_{0.10}(NaCl)_{0.16}$	120.1 (120.1)	17.69 (17.69)	1.00 (1.00)

† Total $\psi = \psi_s + \psi_m$.
‡ Numbers in parentheses are the formula weight, carbon-bound electron, and nitrogen composition of the components, respectively.

Table A-5. Effect of microbial osmoregulation strategy on cell composition under −40 bar and −80 bar NaCl stress.†

Substrate water potential ψt	Turgor pressure ψp	Cell: Intracellular solution: Water	Solutes: Stress: Compound	Stress: Molality M	Stress: Water potential ψss	Stress: Volume V	Compatible: Compound	Compatible: Molality M	Compatible: Water potential ψcs	Compatible: Volume V	Empirical composition: Cell formula $[C_aH_bO_cN_dK_e(NaCl)_g]$	C-bound electrons	C-bound electrons	Formula weight F.W.	Water content ω
bars	bars	g/mole cells		g/1,000 g	bars	ml/mole cells		g/1,000 g	bars	ml/mole cells		e^-eq/mole cells	e^-eq/mole cell N	g/mole cells	g H_2O/g dry cells
Class I—No specific ψ stress-related osmoregulatory mechanism															
−2	+5	241	None	0	0	0	None	0	0	0	$C_{4.0}H_{7.0}O_{1.5}N_{1.0}K_{0.03}$	17.0	17.0	103.5	2.33
−42	+5	237	NaCl	0.90	−40	4	None	0	0	0	$C_{4.0}H_{7.0}O_{1.5}N_{1.0}K_{0.03}(NaCl)_{0.21}$	17.0	17.0	116.0	2.04
−82	+5	233	NaCl	1.78	−80	8	None	0	0	0	$C_{4.0}H_{7.0}O_{1.5}N_{1.0}K_{0.03}(NaCl)_{0.41}$	17.0	17.0	127.0	1.83
Class II—Intracellular accumulation of inducible compatible solutes under water stress															
−2	+5	241	None	0	0	0	None	0	0	0	$C_{4.0}H_{7.0}O_{1.5}N_{1.0}K_{0.03}$	17.0	17.0	103.5	2.33
K^+ glutamate$^-$ as inducible compatible solute															
−42	+5	232	NaCl	0.45	−20	2	Kglut	0.46	−20	7	$C_{4.5}H_{7.9}O_{1.9}N_{1.1}K_{0.14}(NaCl)_{0.10}$	18.9	17.0	129.3	1.79
−82	+5	228	NaCl	1.35	−60	6	Kglut	0.46	−20	7	$C_{4.5}H_{7.8}O_{1.9}N_{1.1}K_{0.14}(NaCl)_{0.31}$	18.9	17.0	140.7	1.62
K^+ glutamate$^-$ and proline as inducible compatible solutes															
−42	+5	223	NaCl	0	0	0	Kglut	0.46	−20	7	$C_{5.4}H_{9.4}O_{2.3}N_{1.3}K_{0.13}(NaCl)_{0.0}$	22.7	17.8	142.8	1.56
							Proline	0.81	−20	11					
−82	+5	220	NaCl	0.90	−40	3	Kglut	0.46	−20	7	$C_{5.4}H_{9.4}O_{2.2}N_{1.3}K_{0.13}(NaCl)_{0.20}$	22.6	17.7	153.7	1.43
							Proline	0.81	−20	11					
K^+ glutamate$^-$ and gamma amino butyrate as inducible compatible solutes															
−42	+5	223	NaCl	0	0	0	Kglut	0.46	−20	7	$C_{5.2}H_{9.4}O_{2.3}N_{1.3}K_{0.13}(NaCl)_{0.0}$	22.0	17.2	140.7	1.58
							GAB	0.81	−20	11					
−82	+5	220	NaCl	0.90	−40	3	Kglut	0.46	−20	7	$C_{5.2}H_{9.4}O_{2.2}N_{1.3}K_{0.13}(NaCl)_{0.20}$	21.9	17.1	151.6	1.45
							GAB	0.81	−20	11					

(continued on next page)

Table A-5. Continued.

		Cell													
		Intracellular solution													
			Solutes												
			Stress				Compatible				Empirical composition				
Substrate water potential ψt	Turgor pressure ψp	Water	Com-pound	Molality M	Water potential ψss	Volume V	Com-pound	Molality M	Water potential ψcs	Volume V	Cell formula $[C_aH_bO_cN_dK_e(NaCl)_g]$	C-bound electrons		Formula weight F.W.	Water content W
bars	bars	g/mole cells		g/ 1,000 g	bars	ml/mole cells		g/ 1,000 g	bars	ml/ cells		e^-eq/ mole cells	e^-eq/ mole cell N	g/ mole cells	g H_2O/ g dry cells
Glycerol as inducible compatible solute															
−42	+5	217	NaCl	0	0	0	Glyc	1.57	−40	24	$C_{3.8}H_{9.7}O_{2.5}N_{1.0}K_{0.03}(NaCl)_{0.0}$	21.7	21.7	134.1	1.62
−82	+5	205	NaCl	0.45	−20	2	Glyc	2.30	−60	34	$C_{5.4}H_{10.7}O_{2.9}N_{1.0}K_{0.03}(NaCl)_{0.09}$	23.5	23.5	151.5	1.35
Class III—Intracellular accumulation of constitutive compatible solutes															
K⁺ glutamate⁻ as constitutive compatible solute															
−2	+25	234	None	0	0	0	Kglut	0.46	−20	7	$C_{4.5}H_{7.9}O_{1.9}N_{1.1}K_{0.14}$	18.9	17.0	123.4	1.90
−42	+5	232	NaCl	0.45	−20	2	Kglut	0.46	−20	7	$C_{4.5}H_{7.9}O_{1.9}N_{1.1}K_{0.14}(NaCl)_{0.10}$	18.9	17.0	129.3	1.79
	+25	230	NaCl	0.90	−40	4	Kglut	0.46	−20	7	$C_{4.5}H_{7.8}O_{1.9}N_{1.1}K_{0.14}$	18.9	17.0	135.1	1.70
−82	+5	228	NaCl	1.35	−60	6	Kglut	0.46	−20	7	$C_{4.5}H_{7.8}O_{1.9}N_{1.1}K_{0.14}(NaCl)_{0.31}$	18.9	17.0	140.7	1.62
	+25	226	NaCl	1.78	−80	8	Kglut	0.46	−20	7	$C_{4.5}H_{7.8}O_{1.9}N_{1.1}K_{0.13}$	18.8	17.1	145.9	1.55
Arabitol as constitutive compatible solute															
−2	+25	223	None	0	0	0	Arab	0.81	−20	18	$C_{4.9}H_{9.1}O_{2.4}N_{1.0}K_{0.03}$	20.9	20.9	130.3	1.71
−42	+5	221	NaCl	0.45	−20	2	Arab	0.81	−20	18	$C_{4.9}H_{9.1}O_{2.4}N_{1.0}K_{0.03}(NaCl)_{0.19}$	20.9	20.9	135.9	1.63
	+25	219	NaCl	0.90	−40	4	Arab	0.81	−20	18	$C_{4.9}H_{9.1}O_{2.3}N_{1.0}K_{0.03}(NaCl)_{0.20}$	20.8	20.8	141.4	1.55
−82	+5	218	NaCl	1.35	−60	5	Arab	0.81	−20	18	$C_{4.9}H_{9.1}O_{2.4}N_{1.0}K_{0.03}(NaCl)_{0.29}$	20.8	20.8	146.7	1.49
	+25	216	NaCl	1.78	−80	7	Arab	0.81	−20	18	$C_{4.9}H_{9.1}O_{2.4}N_{1.0}K_{0.03}(NaCl)_{0.38}$	20.8	20.8	151.8	1.42
Class IV—Intracellular accumulation of inducible and constitutive compatible solutes															
K⁺ glutamate⁻ as constitutive and proline as inducible compatible solutes															
−2	+25	234	None	0	0	0	Kglut	0.46	−20	7	$C_{4.5}H_{7.9}O_{1.9}N_{1.1}K_{0.14}$	18.9	17.0	123.4	1.90
−42	+5	223	NaCl	0	0	0	Kglut	0.46	−20	7	$C_{5.4}H_{9.4}O_{2.3}N_{1.3}K_{0.13}(NaCl)_{0.0}$	22.7	17.8	142.8	1.56
							Proline	0.81	−20	11					

(continued on next page)

Table A-5. Continued.

Substrate water potential ψt	Turgor pressure ψp	Water	Stress Com-pound	Stress Molality M	Stress Water potential ψss	Stress Volume V	Compatible Com-pound	Compatible Molality M	Compatible Water potential ψcs	Compatible Volume V	Cell formula $[C_aH_bO_cN_dK_e(NaCl)_g]$	C-bound electrons		Formula weight F.W.	Water content W
bars	bars	g/mole cells		g/1,000 g	bars	ml/mole cells		g/1,000 g	bars	ml/mole cells		e^-eq/mole cells	e^-eq/mole cell N	g/mole cells	g H_2O/g dry cells
	+25	221	NaCl	0.45	−20	2	Kglut	0.46	−20	7	$C_{5.4}H_{9.4}O_{2.3}N_{1.3}K_{0.13}(NaCl)_{0.10}$	22.7	17.7	148.2	1.49
							Proline	0.81	−20	11					
−82	+5	220	NaCl	0.90	−40	3	Kglut	0.46	−20	7	$C_{5.4}H_{9.4}O_{2.2}N_{1.3}K_{0.13}(NaCl)_{0.20}$	22.6	17.7	153.7	1.43
							Proline	0.81	−20	11					
	+25	218	NaCl	1.35	−60	5	Kglut	0.46	−20	7	$C_{5.4}H_{9.4}O_{2.2}N_{1.3}K_{0.13}(NaCl)_{0.29}$	22.6	17.7	158.9	1.37
							Proline	0.81	−20	11					
K^+ glutamate$^-$ as constitutive and glycerol as inducible compatible solutes															
−2	+25	234	None	0	0	0	Kglut	0.46	−20	7	$C_{4.5}H_{7.9}O_{1.9}N_{1.1}K_{0.14}$	18.9	17.0	123.4	1.90
−42	+5	221	NaCl	0	0	0	Kglut	0.46	−20	7	$C_{5.0}H_{9.2}O_{2.4}N_{1.1}K_{0.13}(NaCl)_{0.0}$	21.3	19.3	138.2	1.60
							Glyc	0.80	−20	13					
	+25	211	NaCl	0	0	0	Kglut	0.46	−20	7	$C_{5.5}H_{10.4}O_{2.9}N_{1.1}K_{0.13}(NaCl)_{0.0}$	23.3	21.3	151.2	1.39
							Glyc	1.57	−40	23					
−82	+5	202	NaCl	0	0	0	Kglut	0.46	−20	6	$C_{5.8}H_{11.4}O_{3.2}N_{1.1}K_{0.12}(NaCl)_{0.0}$	25.0	23.2	162.4	1.24
							Glyc	2.30	−60	33					
	+25	201	NaCl	0.45	−20	2	Kglut	0.46	−20	6	$C_{5.8}H_{11.4}O_{3.2}N_{1.1}K_{0.12}(NaCl)_{0.0}$	25.0	23.2	167.3	1.20
							Glyc	2.30	−60	33					
Arabitol as constitutive and glycerol as inducible compatible solutes															
−2	+25	223	None	0	0	0	Arab	0.81	−20	18	$C_{4.9}H_{9.1}O_{2.4}N_{1.0}K_{0.03}$	20.9	20.9	130.3	1.71
−42	+5	211	NaCl	0	0	0	Arab	0.81	−20	18	$C_{5.3}H_{10.4}O_{2.9}N_{1.0}K_{0.03}(NaCl)_{0.0}$	23.0	23.2	144.3	1.46
							Glyc	0.80	−20	12					
	+25	202	NaCl	0	0	0	Arab	0.81	−20	17	$C_{5.7}H_{11.4}O_{3.2}N_{1.0}K_{0.03}(NaCl)_{0.0}$	24.8	25.1	156.2	1.29
							Glyc	1.57	−40	22					

Column groups: Cell — Intracellular solution (Water; Solutes: Stress, Compatible); Empirical composition (Cell formula, C-bound electrons, Formula weight, Water content).

(continued on next page)

Table A-5. Continued.

Substrate water potential ψt	Turgor pressure ψp	Water	Stress: Compound	Stress: Molality M	Stress: Water potential ψss	Stress: Volume V	Compatible: Compound	Compatible: Molality M	Compatible: Water potential ψcs	Compatible: Volume V	Cell formula $[C_aH_bO_cN_dK_e(NaCl)_g]$	C-bound electrons	C-bound electrons	Formula weight F.W.	Water content W
bars	bars	g/mole cells		g/1,000 g	bars	ml/mole cells		g/1,000 g	bars	ml/cells		e^-eq/mole cells	e^-eq/mole cell N	g/mole cells	g H_2O/g dry cells
−82	+5	193	NaCl	0	0	0	Arab	0.81	−20	16	$C_{6.1}H_{12.4}O_{3.6}N_{1.0}K_{0.03}(NaCl)_{0.0}$	26.5	26.8	167.0	1.16
							Glyc	2.30	−60	32					
	+25	192	NaCl	0.45	−20	2	Arab	0.81	−20	16	$C_{6.1}H_{12.3}O_{3.6}N_{1.0}K_{0.02}(NaCl)_{0.09}$	26.4	26.7	171.6	1.12
							Glyc	2.30	−60	31					
K⁺ glutamate⁻ and arabitol as constitutive and glycerol as inducible compatible solutes															
−2	+45	217	None	0	0	0	Kglut	0.46	−20	7	$C_{5.4}H_{9.9}O_{2.8}N_{1.1}K_{0.03}$	22.6	20.5	147.9	1.47
							Arab	0.81	−20	18					
−42	+5	217	NaCl	0	0	0	Kglut	0.46	−20	7	$C_{5.4}H_{9.9}O_{2.8}N_{1.1}K_{0.03}(NaCl)_{0.0}$	22.6	20.5	147.9	1.47
							Arab	0.81	−20	18					
	+45	197	NaCl	0	0	0	Kglut	0.46	−20	6	$C_{6.1}H_{12.0}O_{3.6}N_{1.1}K_{0.11}(NaCl)_{0.0}$	26.3	24.3	171.7	1.15
							Arab	0.81	−20	16					
							Glyc	1.57	−40	22					
−82	+5	197	NaCl	0	0	0	Kglut	0.46	−20	6	$C_{6.1}H_{12.0}O_{3.6}N_{1.1}K_{0.11}(NaCl)_{0.0}$	26.3	24.3	171.7	1.15
							Arab	0.81	−20	16					
							Glyc	1.57	−40	22					
	+45	188	NaCl	0.45	−20	1	Kglut	0.46	−20	6	$C_{6.4}H_{12.8}O_{4.0}N_{1.1}K_{0.11}(NaCl)_{0.08}$	27.7	25.6	185.1	1.01
							Arab	0.81	−20	15					
							Glyc	1.57	−40	31					

† See Table A-4 and text for calculation details and Tables A-1 and A-2 for supplementary data. All systems are assumed to show a constant intracellular solution volume of 234 ml•mole cells⁻¹ and a constant basal solute composition of 0.21 molality with a mean F.W. of 124 g/mole and a ψ of −7 bars; and the cell F.W. includes 9.3 g phosphate, Mg etc./mole cells (Table A-4). The dry weight-based stress or compatible solute composition is given by $C = 0.1M \times M.W. \times \omega$ where C is the composition (g solute•100 g dry weight cells⁻¹), M is the intracellular molality of the solute, M.W. is the molecular weight of the solute, and ω is the cell water content (g H_2O•g dry weight of cells⁻¹ = g H_2O•mole cells⁻¹/cell F.W.).

CLASS I: NO COMPATIBLE SOLUTE PRODUCTION

Microorganisms in this class are potentially the most xerosensitive, particularly to matric ψ stress, since they do not induce compatible solutes under ψ stress nor do they accumulate compatible solutes constitutively.

Under unstressed conditions these microorganisms show a ψ and intracellular solute balance similar to that of the typical Gr^- bacterial cell: An intracellular ψ_s of about -7 bars due exclusively to basal intermediary metabolites, and a ψ_p of $+5$ bars for cells in a -2 bar growth medium (Fig. 1, 2; Tables 3, A-5).

Tolerance to ψ_s Stress—The immediate response of unstressed cells to a ψ downshock of about 5 bars will be loss of turgor accompanied by a relatively small water loss, the magnitude of the latter being dependent on the modulus of elasticity of the cell. The immediate response to a ψ stress in excess of the 5 bar buffer, e.g. -80 bar NaCl (Fig. 2), will be sub-

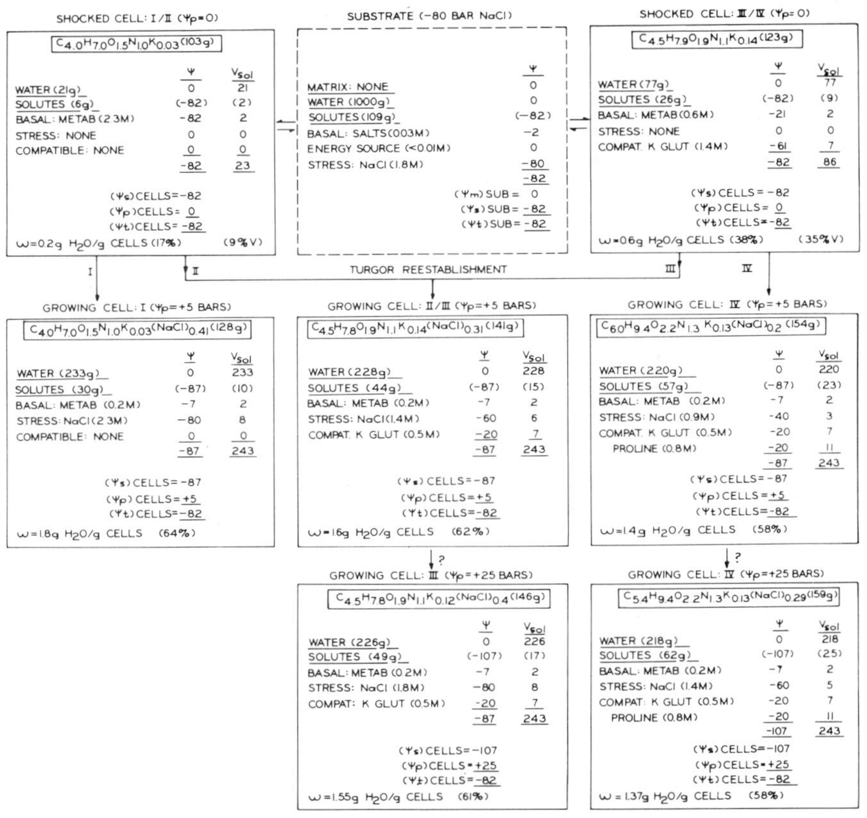

Fig. 2. Schematic of the shock and growth response to a -80 bar NaCl stress of representatives of the major osmoregulation strategy-based classes of microorganisms: Class I (no compatible solutes); Class II (-20 bar K^+ glutamate$^-$ as inducible compatible solute); Class III (-20 bar K^+glutamate$^-$ constitutive); Class IV (-20 K^+glutamate$^-$ constitutive and -20 bar proline inducible). See Appendix II for details and additional data.

Table 3. Compatible solute-based classification of microbial water relations.

Physiological class	Microbial group	Specific representatives†	Compatible solute characteristic: Constitutive		Compatible solute characteristic: Inducible		ψ buffer capacity for	
			Amino acid	Other	Primary	Secondary	Shock	Growth
							bars	bars
CLASS I No compatible solutes	Some Gr^- bacteria	*Spirillum*? (1)	None	None	None	None	5	0 to 5
CLASS II Inducible compatible solutes only	Most Gr^- bacteria	*Pseudomonas* (2); *Beneckea* (3)	None	None	K^+glut$^-$	None	5	20 to 25
		Salmonella, *Klebsiella* (2)	None	None	K^+glut$^-$	Prol/GAB	5	40 to 45
	Some diatoms	*Cyclotella* (4)	None	None	Proline	None	5	20 to 25
	Some algae, Some yeasts, fungi?	*Dunaliella terticola*, *S. cerevisiae* (1)	None	None	Glycerol	None	5	60 to 65
		Platymonas, *Lichina* (4)	None	None	Polyol	None	5	60 to 65
CLASS III Constitutive compatible solutes only	Some Gr^+ bacteria. Streptomycetes?	*Arthrobacter* (5)	K^+glut$^-$	None	None	None	25	0 to 25
	Some lichens. Some free-living fungi, yeasts, algae?	*Peltigera* (6)	None	Mannitol	None	None	25	0 to 25
		?	K^-glut$^-$	Polyol	None	None	45	0 to 45
CLASS IV Constitutive and Inducible Compatible solutes	Most Gr^+ bacteria. Most streptomycetes?	*Streptococcus* (2)	K^+ glut$^-$	None	GAB	None	25	20 to 45
		Staphylococcus (2,7); *Bacillus* (2)	K^+glut$^-$	None	Proline	None	25	20 to 45
	Most yeasts. Most soil fungi, algae?	*Saccharomyces cerevisiae*? (8)	K^+glut$^-$	None	Glycerol	None	25	60 to 85
		Saccharomyces rouxii (1)	None	Arabitol	Glycerol	None	25	60 to 85
		Saccharomyces rouxii?	K^+glut$^-$?	Arabitol	Glycerol	None	45	60 to 105
	Halophilic bacteria. Halophilic streptomycetes?	*Halobacteria* (1)	None	K^+Cl^-	K^+Cl^-	None		
		Halococcus (1)	K^+glut$^-$?	K^+Cl^-	K^+Cl^-	None		
	Halophilic algae	*Dunaliella viridis* (1)	None	Polyol?	Glycerol	None		

† References: 1. Brown, 1976, 1978; 2. Measures, 1975; 3. Makemson and Hastings, 1979; 4. Schobert, 1977; 5. Unluturk, 1977, and Harris, unpublished data; 6. Smith, 1979; 7. Koujima et al., 1978; 8. Tempest et al., 1970.

stantial water loss until ψ equilibrium is achieved by basal metabolite concentration [decrease in $(\psi_{bs})_{cells}$] and, as membrane distortion due to cell dehydration becomes more severe, influx of "permeant" stress solute (NaCl in this case) by passive diffusion [increase in $(\psi_{ss})_{cells}$]. Dehydration will result in cell shrinkage or, for organisms like Gr^- bacteria with relatively weak bonding between the cell membrane and the cell wall, plasmolysis due to protoplast dehydration away from the cell wall. Assuming that such plasmolysis is not so severe as to cause irreversible damage and cell death, and that resumption of growth under −80 bar NaCl ψ stress requires reestablishment of full turgor, the cell must adjust its intracellular ψ by accumulating NaCl such that $(\psi_{ss})_{cell} = (\psi_{ss})_{sub} = -80$ bar NaCl. This is nominally a nonenergy demanding response since no additional osmoregulatory control is involved; reestablishment of 0 rather than the full +5 bar ψ_p would, apart from introducing possible cell division problems related to lack of internal hydrostatic pressure, require osmoregulatory energy for Na^+ efflux to maintain the −80 bar NaCl (out) vs. −75 bar NaCl (in) gradient across the cell membrane.

Ultimate growth rate under ψ_s stress depends on the tolerance of cell enzymes (and solute transport systems) to the specific forms and amounts of stress solutes (nonselectively) accumulated intracellularly from the medium. Thus, the ψ relations of Class I organisms will be dependent on the specific chemical nature of the ψ-controlling stress solutes rather than on ψ_s per se. For example, Na^+ inhibition of intracellular enzymes will be far more severe for −40 bar Na_2SO_4 (1.6 *M* Na^+) than for −40 bar NaCl (0.9 *M* Na^+) than for −40 bar KCl (Table A-2a).

Due to their probable tendency to possess Gr^--like cell walls, Class I organisms are likely to be relatively sensitive to upshock-induced plasmoptys on transfer from ψ stress media to basal (high ψ) media. For example, the immediate response of back transfer of −80 bar NaCl-growing cells to the −2 bar basal medium will be water influx to produce a ψ_p of +85 bars, unless counterbalancing NaCl efflux is rapid enough to prevent attainment of the full ψ_p potential.

Tolerance to ψ_m Stress—The immediate response to ψ downshock will be similar to that shown for ψ downshock due to ψ_s stress: Turgor loss followed by dehydration (Fig. 2). However, in contrast to ψ_s stress, for ψ_m stress, there are no stress solutes available for subsequent intracellular ψ adjustment. Thus, these organisms are highly sensitive to ψ_m stress: The maximum tolerance for growth is −5 bars (assuming growth is possible for $\psi_p = 0$); if full turgor ($\psi_p = +5$) is required for maximum growth rate, a ψ_m stress of only a few bars will cause at least a reduction in growth rate if not growth cessation.

Representative Microorganisms—Unequivocal data identifying specific representative Class I microorganisms are not available. The extreme NaCl sensitivity of certain fresh water *Spirillum* species would make them likely candidates (Table 3). However, the converse is not necessarily true: Moderately salt-tolerant organisms cannot be excluded from this class since the intracellular enzymes of many organisms are not highly salt-sensitive (Measures, 1975; Brown, 1976). Class I organisms are not suited ecologically for competitive growth in soil because of their extreme sensitivity to matric ψ stress.

CLASS II: INDUCIBLE COMPATIBLE SOLUTES ONLY

In the absence of ψ stress, Class II organisms show a ψ balance and cell composition similar to that shown by Class I organisms (Tables 3, A-5; Fig. 2). Under ψ stress, compatible solutes such as amino acids (procaryotes) and polyols (eucaryotes) are induced and accumulated for intracellular ψ control.

Tolerance to ψ_s Stress—Tolerance to the plasmolytic effect of ψ downshock is analogous to Class I organisms: A 5 bar buffer before the onset of severe dehydration. However, unlike Class I organisms, where turgor reestablishment can be accomplished only by stress solute influx, turgor reestablishment for Class II organisms is potentially a function of both stress solute influx and induction and intracellular accumulation of compatible solutes. The relative importance of the latter is illustrated by the acceleration of cell rehydration, as measured by refractivity, in the presence as compared to the absence of a carbon/energy source (Bovell et al., 1963; Knowles and Smith, 1971).

The nature and level of inducible compatible solute production is organism-dependent, with ψ buffer capacities ranging from about -20 bars to -40 bars for procaryotic induction of amino acids such as K^+ glutamate$^-$, proline and GAB and over -60 bars for eucaryotic induction of polyols such as glycerol (Brown, 1976, 1978; Measures, 1975; Makemson and Hastings, 1979; Tables 3, A-5). Growth under ψ_s stress within the inducible compatible solute buffer capacity will tend to be relatively independent of the specific nature of the ψ_s stress solutes and should reflect more the tolerance of cell enzymes to the concentration of the specific compatible solute(s) accumulated for intracellular ψ adjustment. Cell yield should reflect the increased osmoregulatory energy cost of stress solute exclusion and compatible solute retention. In practice, it is likely that stress solute exclusion is not absolute even at low ψ_s stress levels well within the ψ buffer capacity of the induced compatible solutes (Measures, 1975). The relative extent of incomplete exclusion of stress solutes is probably organism-dependent, as is the stress solute sensitivity of intracellular enzymes (Measures, 1975). According to Brown (1976, 1978) the variation in stress solute (e.g. NaCl) tolerance of key intracellular enzymes from a restricted range of non-halophilic organisms differing widely in xerotolerance was relatively small and played a minor role in explaining the observed differences in xerotolerance.

Plasmoptys tolerance to upshock is analogous to Class I organisms from the standpoint that both groups are probably characterized by relatively weak cell walls. However, because of higher ψ_s stress growth tolerance due to the presence of induced compatible solutes, the maximum transient ψ_p potential will tend to be higher for Class II vs. Class I organisms. Resistance to plasmoptys will depend on cell wall strength and the relative rates of water influx vs. stress and compatible solute efflux.

Tolerance to ψ_m Stress—The immediate downshock response is turgor loss followed by cell dehydration with a 5 bar buffer capacity (Fig. 2). However, since stress solute influx is not possible, plasmolysis by ψ_m

downshock ($\psi < -5$ bars) is potentially much more severe than for a ψ_s downshock of similar magnitude. Subsequent turgor reestablishment of dehydrated cells is accomplished by induction of compatible solutes, an option unavailable to Class I organisms. Ultimate growth tolerance to ψ_m stress is limited by the level of inducible compatible solute production (Tables 3, A-5), although the mechanism of cell wall ψ adjustment will also play a major role in determining growth tolerance.

Representative Microorganisms—Most Gr^- bacteria probably belong to this class. Typical organisms accumulating only K^+ glutamate$^-$ include *Pseudomonas aeruginosa* and *Vibrio parahaemolyticus* (Measures, 1975), *Escherichia coli* (Wasileski and Harris, 1979), and *Beneckea harveyi* (Makemson and Hastings, 1979); organisms capable of supplementing K^+ glutamate$^-$ with proline and/or GAB accumulation under ψ_s stress include diverse Gr^- bacteria representing *Salmonella*, *Lactobacillus*, *Serratia* and *Klebsiella* (*Aerobacter*) (Measures, 1975; Brown, 1976; Tempest et al., 1970).

Eucaryotes, as represented by low to moderately xerotolerant *Saccharomyces* yeasts and algae such as *Dunaliella tertiolecta*, characteristically induce and accumulate glycerol for intracellular ψ control under stress (Brown, 1978). Due to the relatively high permeability of microbial membranes to nonionic solutes like glycerol, induction of glycerol as a compatible solute is accompanied by substantial leakage loss of glycerol into the surrounding medium (Brown, 1978). Physiological control mechanisms for glycerol induction under ψ stress are reviewed by Brown (1978).

As summarized by Schobert (1977), under salt stress the green alga *Platymonas subcordiformis* induces mannitol, the marine blue-green algae lichen *Lichina pygmaea* induces mannosidomannitol, the chrysophyte flagellate *Monochrysis lutheri* induces cyclohexanol, and certain red and golden brown fresh water algae induce the galactosylglycerides, floridoside and isofloridoside. In addition, under salt stress the diatom *Cyclotella meneghiniana* induces proline (Schobert, 1977), a common osmoregulator induced by various higher plants (Hsaio et al., 1976). However, whether some or all of these photosynthetic eucaryotes belong in Class IV rather than in Class II requires quantitative knowledge of the intracellular solute composition of the unstressed cells in addition to information on changes in intracellular composition under ψ stress.

Low to moderately xerotolerant soil and aquatic fungi with relatively fragile cell walls, such as *Phytopthora* and other phycomycete fungi (Kouyeas, 1964; Sommers et al., 1970; Griffin, 1978) are probable members of Class II, although experimental proof is not yet available.

CLASS III. CONSTITUTIVE COMPATIBLE SOLUTE PRODUCTION ONLY

This class accumulates compatible solutes such as K^+ glutamate$^-$ (procaryotes) and arabitol, erythritol, mannitol, and/or K^+ glutamate$^-$ (eucaryotes) constitutively but is incapable of supplementary inducible compatible solute production under ψ stress (Table 3, A-5; Fig. 1, 2). The

ψ and solute balance of unstressed cells reflects the presence of the constitutive compatible solute, as illustrated by the typical Gr^+ bacterial cell in Fig. 1. The high turgor characteristic of unstressed cells of this class imposes an associated characteristic requirement for strong cell walls.

Tolerance to ψ_s Stress—The ψ downshock buffer capacity is dependent on the level of constitutive compatible solute: For example, a 20 bar level of constitutive compatible solute gives a downshock buffer capacity of about 25 bars. A downshock ψ stress in excess of the buffer capacity will cause cell dehydration, counterbalanced ultimately by stress solute influx. However, Class III organisms are far less sensitive than Class I and II organisms to the potential plasmolytic effect of cell dehydration under ψ stress in excess of the buffer capacity, because water loss to achieve ψ equilibrium by intracellular solute concentration is substantially decreased by the presence of constitutive compatible solutes. For example, under -80 bar ψ stress, reduction in the concentration of basal metabolites in the typical Gr^- cell (Fig. 1) would require cell dehydration to a water content of $2.3 \times 7/82 = 0.2$ g H_2O/g cells, whereas the same ψ stress would only require dehydration of the Gr^+ cell to a water content of $1.9 \times 27/82 = 0.6$ g H_2O/g cells (Fig. 2). Further protection against plasmolysis under downshock ψ stress results from the characteristically strong binding between the cell wall and membranes shown by Gr^+ bacteria and thick-walled eucaryotic yeasts. Morphologically, the dehydration downshock response of such organisms results in whole cell contraction rather than protoplast plasmolysis (Brown, 1978; Rose, 1976).

Full turgor reestablishment by Class III organisms requires uptake of stress solutes, the extent of uptake being directly dependent on the magnitude of the ψ stress. Such a mechanism is energy conserving from an osmoregulatory cost standpoint but offers no intracellular enzyme protection from the stress solute. This mechanism is consistent with a constitutive compatible solute role emphasizing attenuation of ψ perturbation shock rather than steady state enzyme protection under ψ stress and is characterized, as for Class I organisms, by a specific solute rather than a ψ per se growth response to ψ stress.

Turgor pressure modulation in response to the ψ stress gives a growth ψ buffer capacity dependent on the level of constitutive compatible solute and the optimum ψ_p range for growth (e.g. a 20 bar buffer capacity for a -20 bar constitutive compatible solute level and a minimum ψ_p for optimum growth of $+5$ bars: Fig. 2; Table A-5). Such a mechanism would identify both a downshock and growth protection role for constitutive compatible solutes. Osmoregulatory energy would be needed for control of the stress solute gradient (high out, low in). For a ψ_p of $+5$ bars under < -20 bar ψ stress, the steady state ψ and solute balance of a Class III organism would be identical to that of a Class II organism capable of inducing comparable levels of compatible solutes (Fig. 2; Table A-5). Group III organisms with ψ_p modulation capability would show a growth response to ψ stress within the compatible solute buffer capacity that would be a function more of ψ per se than the specific nature of the ψ stress solute. If growth rate is not dependent on ψ_p over, for example, the $+5$ to $+25$ bar range for the typical Gr^+ cell (Fig. 1), then growth rate would tend to be unaffected by ψ stress down to -20 bars. However, if the optimum ψ_p

was less than the maximum, e.g. if ψ_p optimum was +5 bars, then the optimum ψ for growth would be −22 bars rather than the −2 bars of the basal medium.

The cell composition of Class III organisms will depend not only on the nature and severity of the ψ stress and the nature and level of the constitutive compatible solutes but also on the extent of turgor pressure modulation under ψ stress (Fig. 2; Table A-5).

The upshock tolerance of cells transferred from a low to a high ψ environment is relatively very high because of the strong cell walls shown by this class of organisms.

Tolerance to ψ_m Stress—The downshock tolerance is analogous to that shown for ψ_s stress, about 25 bars buffer capacity. Similarly, the intracellular presence of constitutive compatible solutes and the strong cell wall/membrane properties characteristic of Class III organisms attenuates substantially the plasmolytic effect of severe matric downshock stress.

Turgor reestablishment under impermeant solute/matric stress is impossible because of lack of stress solutes and the inability of Class III organisms to induce compatible solutes. The maximum ψ tolerance for growth is thus dictated by the level of constitutive compatible solute produced and the minimum ψ_p required for growth. Since there are no stress solutes available for uptake, a decrease in ψ_m inevitably causes a concommitant decrease in ψ_p.

No counteracting passive influx of stress solutes can occur to mask the possible presence of an optimum ψ_p < the maximum ψ_p. As for all organisms, prediction of the specific effect of impermeant solute/matrix controlled ψ_m stress on growth necessitates consideration of the mechanism of cell wall ψ adjustment. According to McAneney et al. (1980) Gr^+ cell walls are less likely to undergo severe solute transport-inhibiting structural distortion than Gr^- cell walls under impermeant solute/matric ψ stress.

Representative Microorganisms—Little data are available on this physiological class. The ecological rationale for free-living organisms to develop the capability of producing constitutive but not inducible compatible solutes as well is not readily obvious except in terms of extremely energy-deficient environments subject to frequent ψ_m perturbation where ψ stress survival options must be tightly coupled to energy conservation requirements. Strong cell walls to withstand the high turgor generated in dilute media are a general characteristic of Class III organisms. Preliminary data for the Gr^+ soil organism *Arthrobacter crystallopoietes* show high constitutive levels of glutamate$^-$ and no change in amino acid composition under ψ stress (Wasileski and Harris, unpublished data). Persuasive, but less direct, evidence supports assignment of *A. crystallopoietes* as a Class III organism. The growth response to cation concentration rather than ψ per se, and the relatively unchanged cell composition (C-bound e^-/N ratio) and maintenance energy cost under ψ stress were consistent with an energy-conserving mechanism involving passive stress solute equilibration across the cell membrane. Comparable batch experiments under impermeant solute ψ stress controlled by PEG 4000 (McAneney et al., 1980) identified a slight optimum growth rate at −5 bars and no effect on growth over the −15 bar range evaluated. This response is consistent, particularly when considered in relation to the solute ψ relations of this organism (Wasileski et al., 1978), with a constitutive compatible solute ψ buffer capacity greater than 15 bars (e.g. 20 bars, Fig. 2, Table A-5) and an optimum ψ_p less than the ψ_p maximum shown in basal medium (McAneney et al., 1980).

The "physiological buffering" explanation of why lichens have high photosynthetic rates but very low primary productivity (Farrar, 1976; Smith, 1979) is consistent with the classification of lichens as Class III organisms with polyols as constitutive compatible solutes. As reviewed by Smith (1979), the massive flow of photosynthetically-produced carbohydrate from the alga to the fungus symbiont, independent of ambient ψ conditions, is best explained in terms of its function in maintaining high polyol (arabitol or mannitol) concentration in the fungal cytoplasm, essential for the lichen to survive the continual wetting and drying cycles characteristic of lichen habitats.

CLASS IV. INDUCIBLE AND CONSTITUTIVE COMPATIBLE SOLUTE PRODUCTION

Class IV organisms accumulate compatible solutes such as K^+ glutamate$^-$ (procaryotes and possibly some eucaryotes) and/or polyols such as arabitol, mannitol or erythritol (eucaryotes) constitutively, and under ψ stress induce production and intracellular accumulation of uncharged amino acids such as proline (procaryotes and certain photosynthetic eucaryotes) and/or gamma amino butyrate (procaryotes), and polyols such as glycerol (eucaryotes).

Tolerance to ψ_s Stress—As for Class III organisms, downshock ψ stress tolerance is dictated by the buffer capacity of the constitutive compatible solute (Tables 3, A-5; Fig. 2), and plasmolysis under severe ψ stress is attenuated by the presence of the compatible solute and strong cell wall/membrane properties. Turgor reestablishment under ψ_s stress is facilitated, as for Class II organisms, by induction of compatible solutes rather than being dependent solely on stress solute influx. The ψ buffer capacity for growth is thus a combined function of the concentrations of constitutive and inducible compatible solutes and can reach a substantial level (e.g. up to about 100 bars, Tables 3, A-5). Accordingly, growth response to ψ stress tends to be a function of ψ per se rather than the specific nature of the stress solute over a much broader range of ψ stress conditions than shown for the other groups. Highly xerotolerant (nonhalophilic) microorganisms accumulating high levels (e.g. −40 bars) of constitutive compatible solutes would require exceptionally strong cell walls to withstand the associated high ψ_p (e.g. +45 bars) in basal media. Such organisms might well show optimum ψ for growth substantially lower than that of the basal medium, reflecting an optimum turgor for growth in the +5 to +10 rather than +40 to +50 bar range. Within this context, the relatively very low minimum ψ for growth and retention of cell structure integrity shown by halophiles (Brown, 1978) might reflect, in part, constitutive accumulation of high levels of K^+Cl^- (e.g. −100 bars) as compatible solute, thereby generating destructive turgor pressures even for strong cell-walled halophiles in moderately low ψ media. From a broader standpoint, the magnitude of the upshock (plasmoptys) and downshock (plasmolysis) stresses generated during ψ perturbation within the ψ growth range of Group IV organisms explains the common requirement for stepwise transition between ψ growth tolerance extremes for many xerotolerant microorganisms (Brown, 1978).

Tolerance to ψ_m Stress—Downshock tolerance is analogous to Class III and turgor reestablishment to that of Class II organisms (Fig. 2). Excluding substrate water content-related solute transport problems for matric systems, the maximum growth tolerance to impermeant solute/matric-controlled ψ stress is about -100 bars (Table A-5). Organisms capable of growth at -100 bar ψ_m (Table 2) must either be capable of producing higher levels of compatible solutes than the boundary maxima assumed (Tables 3, A-5) or be capable of modifying their microenvironment to create a "stress" solute substrate ψ component (e.g. cellulase-catalyzed glucose production from plant residues). Alternatively, assumption of a $(\psi_m)_{cell}$ component under conditions of positive turgor would also satisfy cellular growth needs and the requirement for attainment of the total ψ equilibrium (equation 8). The mechanisms used for cell wall ψ control become increasingly important as determinants of growth tolerance under extreme ψ_m stress.

Representative Microorganisms—Most Gr^+ bacteria probably fall into this class. Specific examples include sporeforming bacilli and moderately xerotolerant *Micrococcus*, *Sarcina*, and *Staphylococcus* species (Table 3; Measures, 1975; Brown, 1976; Christian and Waltho, 1964; Koujima et al., 1978), which characteristically show K^+ glutamate$^-$ as constitutive compatible solute and induce proline and/or GAB in response to ψ stress. The ψ growth (Table 2) and survival tolerance and Gr^+ cell wall characteristics of streptomycetes make these organisms likely Class IV candidates, although no direct confirmatory data are available. An interesting evolutionary question is whether streptomycetes use amino acids (heterotrophic procaryotes) or polyols (heterotrophic eucaryotes) as inducible compatible solutes.

Xerotolerant eucaryotes, as represented by sugar tolerant *Saccharomyces* yeasts with massive cell walls, accumulate arabitol as constitutive compatible solute and induce glycerol under ψ stress (Brown, 1978). Restricted membrane permeability is a characteristic property of such organisms, both from the standpoint of resistance to stress solute influx and retention of intracellularly-produced constitutive and inducible compatible solutes (Brown, 1978; Rose, 1976). This phenomenon and the physiological basis of glycerol induction and retention are reviewed by Brown (1978). Evidence for constitutive accumulation of glutamate$^-$ as a compatible solute by *Saccharomyces cerevisiae* is provided by Tempest et al. (1970).

The ψ relations of halophilic procaryotes are complicated by specific NaCl requirements for maintenance of cell membrane and intracellular enzyme stability (Brown, 1978; Lanyi, 1978; Dundas, 1977). Maximum, minimum, and optimum ψ for growth are shown; K^+ (Cl^-) is accumulated intracellularly as a compatible solute. However, chemiosmotic energy transduction considerations (Lanyi, 1978) complicate interpretation of the compatible solute vs. energy transduction implications of high K^+ and Na^+ gradients maintained across the membranes of these organisms. Mechanistic explanations of the high salt minimum shown by the halophilic alga *Dunaliella viridis* are not available (Brown, 1978). Such a minimum is consistent with constitutive accumulation of a high level of compatible solutes, but supporting data are not available (Brown, 1978). *Dunaliella viridis* accumulates glycerol as an inducible compatible solute under ψ stress but the presence of additional intracellular solutes is necessary to explain ψ equilibrium attainment (Brown, 1978).

WATER RELATIONS OF MICROBIAL GROWTH, SURVIVAL AND ACTIVITY

The osmoregulation-based concept of microbial water relations provides a mechanistic rationale for qualification and modification of Scott's (1957) widely accepted generalizations: that as the ψ is reduced below an optimum level, there tends to be an increase in the lag phase of microbial growth, a decrease in the rate of growth, and a decrease in the amount of cell biomass synthesized; that all microorganisms are likely to have a characteristic optimum ψ at which growth occurs most rapidly; that for many microorganisms the biological response to a particular ψ is largely independent of the type of solutes and the total water content of the substrate; and that the more favorable the nutritional conditions for growth, the higher the ψ tolerance for growth.

Effect of Water Potential on Growth

The length of the lag phase following ψ downshock is a function of the severity of the downshock, the level of intracellular constitutive compatible solutes, and the interrelated strength of the cell wall/membrane complex of the microbial cell, the availability of stress solutes to counterbalance dehydration, and the potential for induction of compatible solutes to attenuate the inhibitory effect of stress solutes during turgor reestablishment for growth initiation. As reviewed in the previous section, Class I organisms are the most susceptible to downshock, particularly downshock by matric mechanisms, whereas Class IV organisms with a high ψ buffer capacity (due to the presence of more than one constitutive compatible solute) combined with the associated property of extremely strong cell wall/membrane systems and with a supplementary potential for induction of compatible solutes under ψ stress, show minimal lag following downshock.

Upshock tolerance is a function of the severity of the upshock and cell wall strength. Thus, the constitutive compatible solute-accumulating organisms (Class III and IV), by virtue of their inherently stronger walls than Class I and II organisms, show greater tolerance to plasmoptys and correspondingly shorter lags, under comparable upshock conditions.

Growth rate under ψ stress is a function of the nature and severity of the ψ stress, the intracellular levels of compatible and/or stress solutes, the inherent sensitivity of cell enzymes to stress and compatible solutes, and the equilibrium turgor pressure for growth.

THEORETICAL CONSIDERATIONS OF THE EFFECT OF Ψ STRESS ON MICROBIAL GROWTH RATE AND POPULATION DENSITY

Interpretation and prediction of the effect of ψ stress on microbial growth requires an understanding of the mode of action of the ψ stress on the physiological determinants of growth.

Basic Equations Defining the Determinants of Microbial Growth Under Saturating and Subsaturating Substrate Conditions—This topic is reviewed, with emphasis on the energy-limiting and solute diffusion-restricted conditions characteristic of soil environments, in Harris and Gardner (1980) and Gardner and Harris (1980). In summary, recognizing that the primary physiological determinants of microbial growth rate are the specific rate of assimilatory substrate consumption for biosynthesis of essential cell materials (q_{a-s}^{es}), the specific rate of ATP generation (q_{ATPgen}), and the specific rate of substrate uptake (v_s), the rate of microbial growth ($R_x = dX/dt$) is given by (Harris and Gardner, 1980):

$$R_x = \frac{R_{sq}}{A_s} = \frac{R_{sq}}{A_s^m + A_s^b} = \frac{R_{sq}}{q_s^m/\mu + A_s^b} = R_{sq}Y_s \qquad [11.1]$$

$$= \frac{R_{sq}^b}{A_s^b} = \frac{R_{sq} - R_{sq}^m}{A_{d-s}^b + A_{a-s}} = (R_{sq} - R_{sq}^m)Y_s^b \qquad [11.2]$$

Assuming first order kinetics, Eq. 11 expands to:

$$R_x = \frac{dX}{dt} = \mu X = q_s Y_s X = (q_s - q_s^m)Y_s^b X \qquad [12.1]$$

$$= \frac{\left(\dfrac{q_{ATPgen}}{\xi_{ATP/s}} - q_s^m + q_{a-s}^{es} + q_{a-s}^{st}\right)X}{A_s^b}$$

$$= \frac{\left(\dfrac{q_{ATP} - q_{ATP}^m}{\xi_{ATP/s}\,Ec} + q_{a-s}^{es} + q_{a-s}^{st}\right)X}{\dfrac{A_{ATP}^b}{\xi_{ATP/s}\,Ec} + A_{a-s}^{es} + A_{a-s}^{st}} \qquad [12.2]$$

$$= \frac{(v_s - j_s - q_s^m)X}{A_s^b} = \frac{\left(v_s^{max}\dfrac{S}{S + K_{s(v)}} - j_s - \dfrac{q_{ATP}^m}{\xi_{ATP/s}\,Ec}\right)X}{\dfrac{A_{ATP}^b}{\xi_{ATP/s}\,Ec} + A_{a-s}^{es} + A_{a-s}^{st}} \qquad [12.3]$$

For R_x limited by R_{sf}:

$$R_x = \frac{R_{sf} - q_s^m X}{A_s^b} = \frac{R_{sf} - \left(\frac{q_{ATP}^m}{\xi_{ATP/s}\,Ec}\right) X}{\frac{A_{ATP}^b}{\xi_{ATP/s}\,Ec} + A_{a-s}^{es}} \qquad [12.4]$$

Other relevant relationships are:

$$\frac{1}{Y_s^{\mu max}} = A_s^{\mu max} = A_s^{m(\mu max)} + A_s^b = \frac{q_s^m}{\mu_{max}} + A_{d-s}^b + A_{a-s}$$

$$= \frac{q_{ATP}^m}{\mu_{max}\xi_{ATP/s}\,Ec} + \frac{A_{ATP}^b}{\xi_{ATP/s}\,Ec} + A_{a-s}^{es} + A_{a-s}^{st} \qquad [13.1]$$

$$\frac{1}{Y_s^b} = A_s^b = \frac{A_{ATP}^b}{\xi_{ATP/s}\,Ec} + A_{a-s}^{es} + A_{a-s}^{st} \qquad [13.2]$$

$$q_s^m = \frac{q_{ATP}^m}{\xi_{ATP/s}\,Ec} = \frac{q_{ATP}^{m-v} + q_{ATP}^{m-b}}{\xi_{ATP/s}\,Ec} \qquad [14.1]$$

$$q_s^{\mu max(MIN)} = q_{d-s}^{\mu max(MIN)} + q_{a-s}^{\mu max(MIN)} = \frac{q_{ATPgen}^{\mu max(MIN)}}{\xi_{ATP/s}^{MAX}} + q_{a-s}^{\mu max(MIN)} \qquad [14.2]$$

$$S_{min}^{\mu max} = \frac{(q_s^{\mu max(MIN)} + j_s)K_{s(v)}}{v_s^{max} - (q_s^{\mu max(MIN)} + j_s)} = \frac{v_s^{\mu max(MIN)}\,K_{s(v)}}{v_s^{max} - v_s^{\mu max(MIN)}} \qquad [15.1]$$

$$S_m = \frac{(q_s^m + j_s)K_{s(v)}}{v_s^{max} - (q_s^m + j_s)} \qquad [15.2]$$

$$X_t = X_o \exp(\mu t) = (S_o - S_t)Y_s + X_o \qquad [16.1]$$

$$X_{max(s)} = S_o Y_s + X_o = \frac{S_o}{A_{a-s} + A_{d-s}^b + \frac{q_s^m}{\mu}} + X_o$$

[16.2]

$$= \frac{S_o}{A_{a-s}^{es} + A_{a-s}^{st} + \frac{A_{ATP}^b + q_{ATP}^m/\mu}{\xi_{ATP/s}\,Ec}} + X_o \qquad [16.2]$$

$$N_{x(max)} = \frac{X_{max(s)}}{\chi} \qquad [16.3]$$

$R_{sq} = (dS/dt)_q$ is the total rate of substrate (S) consumption where S, unless specified otherwise, is the electron donor/C source; R_{sq}^{b} and $R_{sq}^{m} = R_{d-sq}^{m}$ are the rates of substrate consumption for biosynthesis and maintenance, respectively; R_{sq}^{b} is composed of R_{d-sq}^{b}, the dissimilatory rate of substrate consumption for energy to drive the biosynthesis reactions, plus R_{a-sq}, the assimilatory rate of substrate consumption for biosynthesis; R_{a-sq} in turn is composed of R_{a-sq}^{es}, the assimilatory rate of substrate consumption for biosynthesis of essential materials, and R_{a-sq}^{st}, the assimilatory rate for biosynthesis of storage and/or ψ stress-related intracelluar materials, R_{sq}^{m} is composed of R_{sq}^{m-v}, the dissimilatory rate for substrate consumption for maintenance of cell integrity and viability, plus (for growing cells) R_{sq}^{m-b}, the dissimilatory rate of substrate consumption needed for maintenance of the biosynthesis machinery necessary for cell growth; q_s, q_s^{b}, q_{d-s}^{b}, q_{a-s}, q_{a-s}^{es}, q_{a-s}^{st}, $q_s^{m} = q_{d-s}^{m}$ (commonly called the maintenance coefficient, M_s), q_s^{m-v} (commonly called the specific rate of endogenous metabolism), and q_s^{m-b} are the analogous specific (with respect to unit cell biomass) rates of substrate consumption; A_s, A_s^{b}, A_{d-s}^{b}, A_{a-s}, A_{a-s}^{es}, A_{a-s}^{st}, $A_s^{m} = A_{d-s}^{m}$, A_s^{m-v} and A_s^{m-b} are the analogous specific amounts of substrate required for biosynthesis and maintenance of unit biomass of cells; μ is the specific growth rate and X is the biomass concentration; $Y_s = dX/dS$ is the specific growth yield per unit mass of substrate; Y_s^{b} is the specific growth yield that would be obtained if all the substrate was used for biosynthesis and there was no maintenance requirement (Y_s at infinite μ); $\xi_{ATP/s}$ is the specific amount of ATP generated per unit mass of substrate dissimilated; Ec is the energy coupling efficiency between ATP generation and ATP use expressed as a ratio with respect to the maximum energy coupling efficiency, Ec^{MAX}, unless specified otherwise; q_{ATP} is the rate of ATP use for biosynthesis plus maintenance, q_{ATP}^{m} is the rate of ATP use for maintenance, and A_{ATP}^{b} is the amount of ATP used to drive biosynthesis of new cell materials (these terms include any nonnegotiable ATP reactions serving indispensable thermodynamic overdrive/slip functions); v_s and j_s are the specific rates of substrate uptake and efflux, respectively; v_s^{max} is the maximum specific rate of substrate uptake with respect to substrate concentration (S) for a given physiological cell condition; $K_{s(v)}$ is the half saturation affinity constant for substrate uptake (S when $v_s = 0.5\ v_s^{max}$); R_{sf} is the rate of substrate supply; the "μ_{max}," "MIN," and "MAX" superscripts denote the values shown at μ_{max} and the minimum and maximum possible values of the physiological property, respectively; for example, $q_s^{\mu max(MIN)}$ is the minimum rate of substrate consumption capable of supporting μ_{max}; $S_{min}^{\mu max}$ is the minimum substrate concentration capable of supporting μ_{max}; S_m is the substrate concentration at which q_s is exactly counter-balanced by q_s^{m}, so that $\mu = 0$ and the physiological problem shifts from growth to survival; X_o, X_t, and $X_{max(s)}$ are the biomass concentrations at time zero, time t and maximum biomass (with respect to S), respectively; S_o and S_t are the substrate concentrations at time zero and time t, respectively; $N_{x(max)}$ is the maximum yield of individual cells and χ is the specific cell mass (cell mass per unit cell).

Under saturating S conditions, the maximum growth rate (μ_{max}) is limited by the physiological properties $q_{a-s}^{es(MAX)}$ and/or q_{ATPgen}^{MAX} or v_s^{MAX}. For μ_{max} limited by $q_{a-s}^{es(MAX)}$ and/or q_{ATPgen}^{MAX}, v_s^{MAX} is either not fully expressed or, if it is fully expressed, appropriate adjustments in other properties must be made to satisfy mass balance laws. Using the "sat," "*" and "lim" superscripts to denote saturating S, S = $S_{min}^{\mu max}$ and limiting S conditions, respectively, with decreasing S: $v_s^{max(sat)} \rightarrow v_s^{MAX}$; $j_s^{(sat)} \rightarrow j_s^{MIN}$; $K_{v(s)}^{(sat)} \rightarrow K_{v(s)}^{MIN}$; $q_{ATP}^{m(sat)} \rightarrow q_{ATP}^{m(MAX)}$ (largely dependent on the cost of operating $K_{v(s)}^{MIN}$); $\xi_{ATP/s}^{(sat)} \rightarrow \xi_{ATP/s}^{MAX}$; $Ec^{(sat)} \rightarrow Ec^{MAX}$; A_{a-s}^{es} and A_{ATP}^{b} primarily reflect the cell composition at a given μ, and should thus remain relatively unchanged as a function of S as long as $\mu = \mu_{max}$; $A_{a-s}^{st(sat)} \rightarrow A_{a-s}^{st(MIN)} \cong 0$; $q_{a-s}^{st(sat)} \rightarrow q_{a-s}^{st(MIN)} \cong 0$. At $S_{min}^{\mu max}$ (the minimum S capable of supporting μ_{max}) μ_{max} becomes simultaneously limited by $v_s^{\mu max(MIN)}$ (the minimum v_s capable of supporting μ_{max}) and the physiological μ_{max}-limiting factor, and all uptake and efficiency properties will tend to become optimized. With decreasing S < $S_{min}^{\mu max}$, $\mu < \mu_{max}$ will decline as a function of the effect of limiting S on $v_s^{(lim)}$ (Eq. 12.3), until, at S = S_m, $\mu = 0$ because $v_s^{(lim)} = q_s^{(lim)} = q_s^{m(lim)}$ (assuming $j_s^{(lim)} \cong 0$). For S < S_m, survival rather than growth is the physiological concern, as discussed later.

The preceding growth equations assume essentially no cell death during growth. This is not necessarily always the case, particularly for microorganisms growing under ψ stress. Physiologically-based growth equations modified for cell death are not currently available, but empirical relationships are given by Pirt (1975):

$$R_{xv} = \frac{dX_v}{dt} = \mu^{obs}X_v = (\mu - \gamma)X_v = (2\Theta - 1)\mu X_v \qquad [17]$$

$$X_{v(t)} = X_{v(o)}\exp(\mu - \gamma)t = X_{v(o)}\exp(2\Theta - 1)\mu t \qquad [18]$$

where $R_{xv} = (dX_v)/dt$ is the rate of change of viable biomass; μ^{obs} is the observed specific growth rate of viable cells; γ is the specific death rate of cells prior to cell division; X_v is the biomass of viable cells; Θ is the viability index, which represents the probability that a newly formed cell will be viable rather than dead at birth.

Effect of ψ Stress-Caused Changes in Specific Efficiency and Rate Properties—The growth response of a microorganism of ψ stress is a function of the effect of the stress, as dictated at least in part by the osmoregulatory mechanism used for intracellular ψ control, on the specific efficiency and rate determinants of growth rate (Eq. 11–15, 17) and population density (Eq. 16, 18). The following example illustrates the growth rate (Fig. 3) and efficiency implications of different physiological responses to a −40 bar NaCl ψ stress of a model Gr^-, osmoregulatory Class II, vs. a model Gr^+, Class III organism.

The growth-limiting carbon/e^-donor source is established as glucose, with S_o = 0.2 e^-eq gluc$\cdot l^{-1}$ (8.333 μM = 1.5 g gluc$\cdot l^{-1}$, Harris and Adams, 1979) O_2 is the terminal electron acceptor; the temperature is mesophilic (20 C). The Class II, Gr^- organism is assigned the following characteristics (Harris and Gardner, 1980). Cell composition: $C_4H_7O_{1.5}NK_{0.03}$

(+9%); F.W. = 103.5 g•mole cells^{-1}; 10^{14} (at μ_{max}) to 10^{15} (at $\mu \rightarrow 0$) individual cells•mole cells^{-1} ($\chi = 1.035 \times 10^{-12}$ to 10^{-13} g•cell^{-1}, respectively), derived from assumption of a cell density = 1.2 g•ml^{-1}, cell water content = 70% (Fig. 1), and dimensions of 1.0 × 3.6 μm (at μ_{max}) and 0.6 × 1.0 μm (at $\mu \rightarrow 0$); assimilatory requirement for glucose, A^{es}_{a-gluc} = 17/24 = 0.708 mole gluc•mole cells^{-1}; A^{st}_{a-gluc} = 0. Primary rate determinants: μ_{max} = 0.3 hr^{-1} limited by $q^{es(MAX)}_{a-gluc}$ = 0.213 mole gluc•mole cells^{-1}•hr^{-1}; q^{MAX}_{ATPgen} = 3.88 moles ATP•mole cells^{-1}•hr^{-1}; v^{MAX}_{s} = 0.474 mole gluc•mole cells^{-1}•hr^{-1}. Energy metabolism, for simplicity: $q^{m(sat)}_{ATP} = q^{m(*)}_{ATP} = q^{m(lim)}_{ATP} = q^{m(MAX)}_{ATP}$ = 0.207 mole ATP•mole cells^{-1}•hr^{-1} (0.002 mole ATP•g cells^{-1}•hr^{-1}), composed of q^{m-v}_{ATP} = 0.021 and q^{m-b}_{ATP} = 0.186 mole ATP•mole cells^{-1}•hr^{-1}; $A^{b(sat)}_{ATP} = A^{b(*)}_{ATP} = A^{(lim)}_{ATP} = A^{(MIN)}_{ATP}$ = 9.66 moles ATP•mole cells^{-1}; $\xi^{(sat)}_{ATP/gluc} = \xi^{(*)}_{ATP/gluc} = \xi^{(lim)}_{ATP/gluc} = \xi^{MAX}_{ATP/gluc}$ = 26 moles ATP•mole gluc^{-1}; $Ec^{(sat)} = Ec^{(*)} = Ec^{(lim)} = Ec^{MAX}$ where the absolute value of Ec^{MAX} reflects the ecological ATP overdrive needs of the organism for growth, but is assigned a value of unity for calculation simplicity (for $Ec = Ec^{MAX} = 1$: $q_{ATPgen} = q_{ATP}$; $A_{ATPgen} = A_{ATP}$). Uptake properties, for simplicity: $v^{max(sat)}_{gluc}$ = 0.332 mole gluc•mole cells^{-1}•hr^{-1} (0.7 v^{MAX}_{gluc}); $v^{max(*)}_{gluc} = v^{max(lim)}_{gluc} = v^{MAX}_{gluc}$ = 0.474 mole gluc•mole cells^{-1}•hr^{-1}; $j^{(sat)}_{gluc} = j^{(*)}_{gluc} = j^{(lim)}_{gluc} = j^{MIN}_{gluc} \cong 0$ mole gluc•mole cells^{-1}•hr^{-1}; $K^{(sat)}_{gluc(v)} = K^{(*)}_{gluc(v)} = K^{(lim)}_{gluc(v)} = K^{MIN}_{gluc(v)}$ = 10 μM gluc. With decreasing S, μ is maintained at μ_{max} (Fig. 3) by appropriate readjustment of $v^{max(sat)}_{gluc} \rightarrow v^{MAX}_{gluc}$ until, at $S^{\mu max}_{min}$ = 23 μmoles gluc•l^{-1} (equation 15.1), v^{MAX}_{gluc} is fully expressed at 0.474 mole gluc•mole cells^{-1}•hr^{-1} and μ_{max} is limited simultaneously by $v^{\mu max(MIN)}_{gluc}$ = 0.332 mole gluc•mole cells^{-1}•hr^{-1} and $q^{es(MAX)}_{a-gluc}$ = 0.213 mole gluc•mole cells^{-1}•hr^{-1}. With decreasing $S_m < S < S^{\mu max}_{min}$ (Fig. 3), $\mu < \mu_{max}$ is controlled by $v^{(lim)}_{gluc} < v^{\mu max(MIN)}_{gluc}$ as a function of decreasing S (equation 12.3) until, at $\mu = 0$, $S = S_m$ = 0.17 μM (equation 15.2); X_{max} will decline in accordance with the increased maintenance energy requirement caused by the decreased μ (Eq. 16.2) but this will be counteracted, from a cell number standpoint, by the associated decline in cell size (i.e., decrease in χ from $10^{-12} \rightarrow 10^{-13}$ g•cell^{-1}) as μ declines from $\mu_{max} \rightarrow$ zero (Eq. 16.3).

Under −40 bar NaCl ψ stress the Gr$^-$ organism accumulates −20 bar K^+ glutamate$^-$ (0.46 *M* intracellular concentration) and −20 bar NaCl (0.45*M*) to achieve ψ equilibrium, with an associated change in cell composition to $C_{4.54}\,H_{7.85}\,O_{1.92}\,N_{1.11}\,K_{0.14}\,(NaCl)_{0.10}$ (+ 7%), F.W. = 129.3 g•mole cells^{-1} (Table A-5). Accordingly, the assimilatory requirement for glucose increases from A_{a-gluc} = 17/24 = 0.708 mole gluc•mole cells^{-1} (unstressed) to A_{a-gluc} = 18.9/24 = 0.788 mole gluc•mole cells^{-1} under ψstress. The inevitable change in cell composition under ψ stress, apart from the growth implications of the associated change in assimilatory requirement (Eq. 11–18), raises serious, but commonly ignored, questions regarding the most appropriate biomass reference for standardization of substrate efficiency and rate properties. Mechanistically, microorganisms operate on a whole cell basis, but changes in cell volume caused by ψ stress-related reductions in μ (apart from any ψ stress per se effects on cell morphology) complicate use of per cell as a biomass parameter. Of the weight-based biomass parameters, cell dry weight, the most commonly used, is the most questionable since it reflects varying degrees of stress solute (e.g. salt) and inducible compatible solute accumulation,

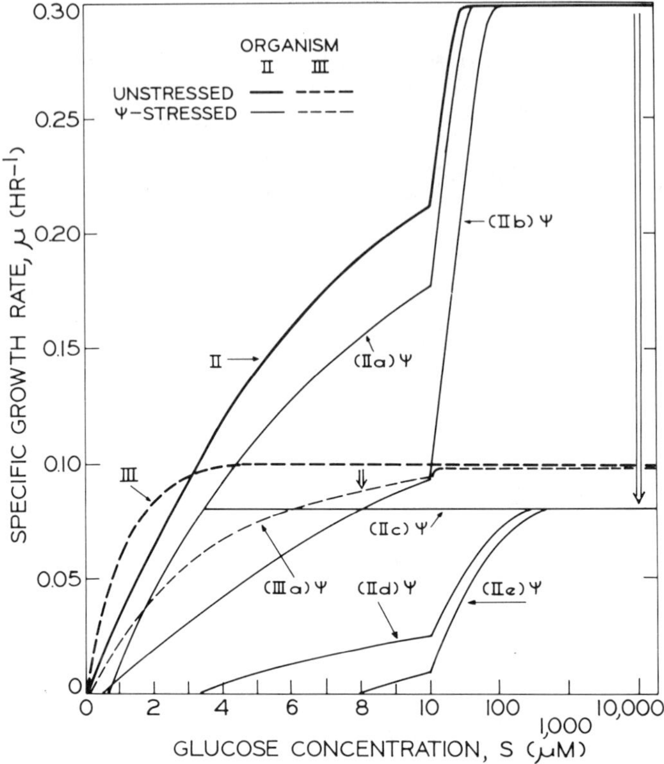

Fig. 3. Effect of a −40 bar NaCl ψ stress on the growth rate of a model osmoregulatory Class II vs. a model osmoregulatory Class III organism, under aerobic saturating and subsaturating glucose growth conditions (see text for details and explanation of terms). The relevant specific growth rate equations are:

$$\mu = q_{gluc}Y_{gluc} = (q_{gluc} - q^{m}_{gluc})Y^{b}_{gluc}$$

For $S \geq S^{\mu max}_{min}$, $\mu = \mu_{max}$ is limited by $q^{es(MAX)}_{a\text{-}gluc}$ or q^{MAX}_{ATPgen}:

$$\mu_{max} = \frac{\dfrac{q_{ATPgen}}{\xi_{ATP/gluc}} - \dfrac{q^{m}_{ATP}}{\xi_{ATP/gluc}Ec} + q^{es}_{a\text{-}gluc} + q^{st}_{a\text{-}gluc}}{\dfrac{A^{b}_{ATP}}{\xi_{ATP/gluc}Ec} + A^{es}_{a\text{-}gluc} + A^{st}_{a\text{-}gluc}}$$

For $s < S^{\mu max}_{min}$, $\mu < \mu_{max}$ is limited by v^{MAX}_{gluc}:

$$\mu = \frac{v^{MAX}_{gluc}\dfrac{S}{S + K_{gluc(v)}} - j_{gluc} - \dfrac{q^{m}_{ATP}}{\xi_{ATP/gluc}Ec}}{\dfrac{A^{b}_{ATP}}{\xi_{ATP/gluc}Ec} + A^{es}_{a\text{-}gluc} + A^{st}_{a\text{-}gluc}}$$

All concentrations are expressed as moles, and the biomass reference, X, is a mole equivalent of cells. The physiological properties of the unstressed Class II organism are:

μ_{max} = 0.3 hr^{-1}, limited by $q_{a\text{-}gluc}^{es(MAX)}$ = 0.213 gluc•X^{-1}•hr^{-1}; q_{ATPgen}^{MAX} = 3.88 ATP•X^{-1}•hr^{-1}; v_{gluc}^{MAX} = 0.474 gluc•X^{-1}•hr^{-1}; q_{ATP}^{m} = 0.207 ATP•X^{-1}•hr^{-1}; $\xi_{ATP/gluc}$ = 26 ATP•$gluc^{-1}$; Ec = 1; $A_{a\text{-}gluc}^{es}$ = 0.708 gluc•X^{-1}; $A_{a\text{-}gluc}^{st}$ = 0.0 gluc•X^{-1}; $K_{gluc(v)}$ = 10 μM; j_{gluc} ≅ 0.0 gluc•X^{-1}•hr^{-1}; $S_{min}^{\mu max}$ = 23.4 μM gluc; S_m = 0.17 μM gluc. Under ψ stress, $A_{a\text{-}gluc}^{st}$ is increased from 0 to 0.08 gluc•X^{-1} and: q_{ATP}^{m} = 0.822 ATP•X^{-1}•hr^{-1}[(IIa)ψ]; $K_{gluc(v)}$ = 30 μM gluc [(IIb)ψ]; μ_{max} = 0.08 hr^{-1} because $q_{a\text{-}gluc}^{es(MAX)}$ = 0.057 gluc•X^{-1}•hr^{-1} and q_{ATP}^{m} = 0.822 ATP•X^{-1}•hr^{-1} [(IIc)ψ]; μ_{max} = 0.08 hr^{-1} because v_{gluc}^{MAX} = 0.125 gluc•X^{-1}•hr^{-1} and q_{ATP}^{m} = 0.822 ATP•X^{-1}•hr^{-1} [(IId)ψ]; μ_{max} = 0.08 hr^{-1} because $\xi_{ATP/gluc}$Ec = 3.9 ATP•X^{--}•hr^{-1} [(IIe)ψ].

The physiological properties of the unstressed Class III organism are: μ_{max} = 0.1 hr^{-1}, limited by $q_{a\text{-}gluc}^{es(MAX)}$ = 0.079 gluc•X^{-1}•hr^{-1}; q_{ATPgen}^{MAX} = 1.58 ATP•X^{-1}•hr^{-1}; v_{gluc}^{MAX} = 0.151 gluc•X^{-1}•hr^{-1}; q_{ATP}^{m} = 0.617 ATP•X^{-1}•hr^{-1}; $\xi_{ATP/gluc}$ = 38 ATP•$gluc^{-1}$; Ec = 1; $A_{a\text{-}gluc}^{es}$ = 0.708 gluc•X^{-1}•hr^{-1}; $A_{a\text{-}gluc}^{st}$ = 0.08 gluc•X^{-1}•hr^{-1}; $K_{gluc(v)}$ = 1.0 μM gluc; j_{gluc} ≅ 0.0 gluc •X^{-1}•hr^{-1}; $S_{min}^{\mu max}$ = 4.0 μM gluc; S_m = 0.12 μM gluc. Under ψ stress: $K_{gluc(v)}$ = 3.0 μM gluc [(IIIa)ψ].

and this leads to progressively more inflated and mechanistically-misleading growth yield as ψ stress becomes more severe (Fig. 2; Table A-5). For example, assuming that the physiological response of the Class II organism to −40 bar NaCl stress was confined solely to the change in cell composition, the dry weighted-based $Y_{gluc}^{\mu max}$ would change from 93.5 to 109.1 g cells•mole $gluc^{-1}$ (17% increase) under ψ stress, whereas the cell organic carbon, carbon-bound electron equivalent, and cell N-based yields would change from 0.60 to 0.64 mole cell C•mole gluc C^{-1} (6% increase), 0.64 to 0.66 cells•e^{-}eq $gluc^{-1}$ (3% increase), and 70.3 to 72.8 mg cell N•g $gluc^{-1}$ (3% increase), respectively, under ψ stress. Similarly, the dry weight-based q_{gluc}^{m} would change from 0.077 to 0.061 mmole gluc•g $cells^{-1}$•hr^{-1} under ψ stress (21% decrease), whereas the C-bound e^{-}eq-based maintenance coefficient would show a lesser change, from 0.468 to 0.421 mmole gluc•e^{-}eq $cells^{-1}$•hr^{-1} (10% decrease). Y_{gluc}^{b} would increase from 95.8 (unstressed) to 111.6 g cells•mole $gluc^{-1}$ (16.5% increase), but would show less change when expressed on an electron equivalent-biomass basis, 15.7 (unstressed) vs. 16.3 (ψ stressed) e^{-}eq cells•mole $gluc^{-1}$ (4% increase). From a growth rate standpoint, the 0.08 mole gluc•mole $cells^{-1}$ increase in assimilatory requirement due to induced "storage" of glutamate for intracellular ψ control will result in an increased uptake rate of Δv_{gluc} = 0.08 × 0.3 = 0.024 mole gluc•mole $cells^{-1}$•hr^{-1} (increase in $v_{gluc}^{\mu max}$ from 0.332 to 0.356 mole gluc•mole $cells^{-1}$•hr^{-1}) required to maintain μ_{max} at 0.3 hr^{-1}. This is well within the glucose uptake rate buffer capacity of 0.474 − 0.332 = 0.142 mole gluc•mole $cells^{-1}$•hr^{-1} at saturating S and will cause only a relatively minor increase in $S_{min}^{\mu max}$ from 23 to 30 μM gluc (Eq. 15.1) and no change in S_m (Eq. 15.2). Similarly, the ATP generation buffer capacity of 3.88 − 3.105 = 0.775 ATP•mole $cells^{-1}$•hr^{-1} will absorb an increased osmoregulatory ATP requirement for concentration gradient control under ψ stress equivalent to an almost 5-fold increase in the maintenance energy coefficient from 0.207 to 0.207 + 0.775 = 0.982 ATP•mole $cells^{-1}$•hr^{-1}, without a reduction in μ_{max} at saturating S. Taking into consideration the cell composition change, the following physiological properties would be shown for a four-fold increase in osmoregulatory maintenance energy (cell mole and e⁻eq as biomass reference) under ψ stress: q^{m}_{gluc} would increase from 0.468 to 1.685 mmoles gluc•e⁻eq cells^{-1}•hr^{-1}; μ_{max} would be maintained at 0.3 hr^{-1} down to $S^{\mu max}_{min}$ = 40 μM; μ at any given S < $S^{\mu max}_{min}$ would decrease somewhat, and S_m would increase from 0.17 to 0.72 μM; $Y^{\mu max}_{gluc}$ would decrease slightly from 15.4 to 14.9 e⁻eq cells•mole gluc^{-1} and Y^{b}_{gluc} would increase slightly from 15.7 to 16.3 e⁻eq cells•mole gluc^{-1}. A relatively high ψ stress interference with the glucose affinity properties, e.g. over a 50-fold increase in $K_{gluc(v)}$ from 10 (unstressed) to >500 μM gluc, could be tolerated without a decrease in μ_{max} (Eq. 12.3). For a more realistic three-fold increase in $K_{gluc(v)}$ under ψ stress, from 10 to 30 μM gluc, the following properties would be shown; q^{m}_{gluc}, $Y^{\mu max}_{gluc}$, and Y^{b}_{gluc} would change in accordance solely with the change in cell composition under ψ stress; μ (Fig. 3) would be maintained at μ_{max} = 0.3 hr^{-1} down to $S^{\mu max}_{min}$ = 3 × 30 = 90 μM gluc (Eq. 15.1) and would decline thereafter as a function of decreasing S, with growth extinction occurring at S_m = 3 × 0.17 = 0.51 μM gluc (Eq. 15.2).

A reduction in μ_{max} under −40 bar NaCl stress to, for example, 0.08 hr^{-1} (Fig. 3) could be caused by a variety of ψ stress effects on the efficiency and rate determinants of growth, each effect having characteristically different implications with respect to growth yield/population density and competitive growth rate in the subsaturating S range. From the former standpoint, it is critical to recognize that, because of the dependency of growth yield on μ (Eq. 13.1), a μ reduction caused by ψ stress will always be accompanied by a reduction in growth yield even if the efficiency parameters are unaffected by the ψ stress. Accordingly, interpretation of growth yield responses to ψ stress for microbial systems showing concommitant ψ stress-caused reductions in μ requires appropriate correction of the yield data for yield reductions due merely to normal maintenance (Wasileski et al., 1978; Unluturk, 1977). For example, taking into account the change in cell composition, a μ_{max} reduction to 0.08 hr^{-1} caused solely by a 73% inhibition of the intracellular enzymes controlling the rate of biosynthesis ($q^{es(MAX)}_{a-gluc}$ reduced from 0.218 to 0.057 mole gluc•mole cells^{-1}•hr^{-1}) under −40 bar NaCl stress would result in the following response: as discussed previously, q^{m}_{gluc} would decrease from 0.468 to 0.421 mmole gluc•e⁻eq cells^{-1}•hr^{-1}; $Y^{\mu max}_{gluc}$ would decline from 15.4 to 15.0 e⁻eq cells•mole gluc^{-1}, caused solely by the decrease in μ, counteracted somewhat by the cell composition change; Y^{b}_{gluc} would increase from 15.7 to 16.3 e⁻eq cells•mole gluc^{-1} because of the increased cell glutamate content under ψ stress; μ_{max} would be maintained at 0.08 hr^{-1} with decreasing S down to $S^{\mu max}_{min}$ = 2.7 μM gluc (Fig. 3), at which point the μ vs. S relationship would become identical to the non $q^{es(MAX)}_{a-gluc}$ − inhibited system (S_m = 0.17 μM gluc). A similar 73% inhibition of the rate of biosynthesis, coupled with a four-fold increase in osmoregulatory maintenance energy, would show the following characteristics: q^{m}_{gluc} would increase from 0.468 to 1.685 mmoles gluc•e⁻eq cells^{-1}•hr^{-1}; $Y^{\mu max}_{gluc}$ would decrease from 15.4 to 12.1 e⁻eq cells•mole gluc^{-1}, and Y^{b}_{gluc} would increase from 15.7 to 16.3 e⁻eq cells•mole gluc^{-1}; μ_{max} would be maintained at 0.08 hr^{-1} until $S^{\mu max}_{min}$= 3.6 μM gluc, and μ would then decline as a function of S, with S_m increased to 0.7 μM gluc because of the increased maintenance coefficient. Interference with the efficiency of energy coupling between glucose dissimilation and ATP generation could also cause a μ_{max} decline to 0.08 hr^{-1}, e.g. a decrease in $\xi_{ATP/gluc}$ from 26 to 3.9 moles ATP•mole $gluc^{-1}$, combined with a four-fold increase in osmoregulatory maintenance energy, would be characterized as follows: q^{m}_{gluc} would increase from 0.468 to 11.23 mmoles gluc•e^-eq $cells^{-1}$•hr^{-1}; $Y^{\mu max}_{gluc}$ would decrease dramatically from 15.4 to 3.19 e^-eq cells •mole $gluc^{-1}$; in marked contrast to the previous systems, Y^{b}_{gluc} would also decrease dramatically from 15.7 down to 5.8 e^-eq cells•mole $gluc^{-1}$; $v^{\mu max(sat)}_{gluc}$ would be fully expressed at v^{MAX}_{gluc} = 0.474 mole gluc•mole $cells^{-1}$• hr^{-1}; μ_{max} = 0.08 hr^{-1} would be limited by v^{MAX}_{gluc}; with decreasing S, μ would decline in accordance with v_{gluc} dependency on S, but because of the relatively low $K_{gluc(v)}$ value of 10 μM gluc, little change in μ_{max} = 0.08 hr^{-1} would be seen in practice until S declined below about 200 μM (Fig. 3; Eq. 12.3); because of the very large increase in q^{m}_{gluc}, S_m would increase from 0.17 to 8.1 μM gluc (Fig. 3). A μ_{max} reduction to 0.08 hr^{-1} resulting solely from an increased osmoregulatory energy requirement under ψ stress would require an almost 50-fold increase in the maintenance coefficient, from 0.207 to 9.91 moles ATP•mole $cells^{-1}$•hr^{-1}, but this would greatly exceed the assigned q^{MAX}_{ATPgen} property of 3.88 moles ATP•mole $cells^{-1}$•hr^{-1}.

In addition to possible ψ stress effects on the efficiency and rate determinants of growth, ψ stress may also increase the death rate (γ) and/or decrease the viability index (Θ) of growing cells (Eq. 17–18), but current lack of mechanistic information precludes theoretical analysis of this aspect of microbial ψ relations.

For comparative purposes, the unstressed and −40 bar salt-stressed μ vs. S curves for the model osmoregulatory Class III, Gr^+ organism are included in Fig. 3. This organism is assigned the following physiological properties (Harris and Gardner, 1980). Cell composition (Fig. 1, Table A-5): $C_{4.54}H_{7.85}O_{1.92}N_{1.11}$ (+8%); F.W. = 123.4 g•mole $cells^{-1}$; assimilatory requirement for glucose, $A_{a\text{-}gluc}$ = 18.9/24 = 0.788 mole gluc•mole $cells^{-1}$ (including 1.9 e^-eq glutamate•mole $cells^{-1}$ as constitutive compatible solute). Primary rate determinants: μ_{max} = 0.1 hr^{-1} limited by $q^{es(MAX)}_{a\text{-}gluc}$ = 0.079 gluc•mole $cells^{-1}$•hr^{-1}; q^{MAX}_{ATPgen} = 1.979 moles ATP•mole $cells^{-1}$• hr^{-1}; v^{MAX}_{gluc} = 0.151 mole gluc•mole $cells^{-1}$•hr^{-1}. Energy metabolism: q^{m}_{ATP} = 0.617 mole ATP•mole $cells^{-1}$•hr^{-1}, consisting of q^{m-v}_{ATP} = 0.006 and q^{m-b}_{ATP} = 0.601 mole ATP•mole $cells^{-1}$•hr^{-1}; A^{b}_{ATP} = 9.66 moles ATP•mole $cells^{-1}$; $\xi_{ATP/gluc}$ = 38 moles ATP•mole $gluc^{-1}$; Ec = E^{MAX} = unity. Uptake properties: $v^{max(sat)}_{gluc}$ = 0.121 mole gluc•mole $cells^{-1}$•hr^{-1} (0.8 v^{MAX}_{gluc}); j_{gluc} ≅ 0 mole gluc•mole $cells^{-1}$•hr^{-1}; $K^{(sat)}_{gluc(v)}$ = $K^{(*)}_{gluc(v)}$ = $K^{(lim)}_{gluc(v)}$ = $K^{MIN}_{gluc(v)}$ = 1.0 μmole gluc•l^{-1}. With decreasing S, μ is maintained at μ_{max} (Fig. 3) by appropriate readjustment of $v^{max(sat)}_{gluc} \rightarrow v^{MAX}_{gluc}$ until, at $S^{\mu max}_{min}$ = 4 μmoles gluc•l^{-1} (Eq. 16.1), v^{MAX}_{gluc} is fully expressed at 0.151 mole gluc•mole $cells^{-1}$•hr^{-1} and μ_{max} is limited simultaneously by $v^{\mu max(MIN)}_{gluc}$ = 0.121 and $q^{es(MAX)}_{a\text{-}gluc}$ = 0.079 mole gluc•mole $cells^{-1}$•hr^{-1}. Below $S^{\mu max}_{min}$, $\mu < \mu_{max}$ is controlled by the effect of S on v_{gluc} (Eq. 12.3) until, at μ = 0, S = S_m = 0.12 μM gluc (Eq. 15.2). From a competitive growth standpoint, the Gr^+ organism would be outcompeted by the Gr^- organism at saturating S, and over a large range of subsaturating S, but a crossover would be shown at about 3 μM gluc, below which the Gr^+ organism would dominate (Fig. 3). Competitive ability under ψ stress would depend on the relative effect of the ψ stress on the growth determinants of the two organisms. Assumption of a relatively high ψ_s tolerance for the Gr^+ organism such that, for example, the only effect of a -40 bar NaCl ψ stress was a three-fold increase in $K_{gluc(v)}$, would result in no change in $\mu_{max} = 0.1\ hr^{-1}$ in batch culture (Fig. 3), but this would not necessarily exclude a relatively higher competitive ability under subsaturating S, for the ψ-stressed Gr^- (μ_{max} under stress $= 0.08\ hr^{-1}$) vs. the ψ-stressed Gr^+ (μ_{max} under stress $= 0.1\ hr^{-1}$) organism (Fig. 3).

Additional Phenomena Controlling the Effect of Matric ψ Stress on μ

As described above, the effect of ψ stress on growth is a function of the physiological effect of the osmoregulation strategy of the microbe on the efficiency and rate determinants of growth rate and population density (Eq. 11–18). An additional ψ_m per se effect, as yet ill-defined but possibly involving more severe problems of cell wall ψ adjustment under matrix vs. solute ψ stress (McAneney et al., 1979, 1980), may contribute to the effect of ψ_m on growth, independent of whether ψ_m is controlled by impermeant solute or matrix phenomena. The effect of matrix-controlled ψ_m is further compounded as a function of the indirect effect of matrix-controlled ψ_m on the S dependency of growth rate (Eq. 12.3), as dictated by the role of substrate water content as a diffusion-related determinant of solute resupply (R_{sf} in Eq. 12.4) to the cell. Interrelations between solute transport, substrate water content, and matrix-controlled ψ_m are discussed elsewhere (Papendick and Campbell, Chapter 1). The relative importance of osmoregulation strategy, ψ_m per se, and ψ_m/water content-related solute diffusion as determinants of microbial growth response to matric or combined matric and solute ψ stress (as occurs in saline soils and in fertilizer bands) is largely conjectural at this time. The ψ_m per se effect may dominate the growth response of Class I and II organisms with Gr^--like cell walls (McAneney et al., 1979, 1980). Solute diffusion becomes progressively more important with decreasing matrix-controlled ψ_m/water content in accordance with the sorption moisture isotherm characteristics of the substrate (Papendick and Campbell, Chapter 1). However, it should be recognized that under the relatively static conditions prevailing in matric (as compared to liquid) systems solute diffusion commonly becomes the ultimate growth-limiting factor even under high matrix-controlled ψ_m ($\psi_t > -1$ bar) conditions where ψ per se effects on efficiency and uptake are negligible.

EXPERIMENTAL DETERMINATION OF THE EFFECT OF Ψ STRESS ON GROWTH

Experimentally, the effect of ψ stress on μ_{max} may be determined in batch culture (log phase $S \gg K_{s(v)}$) as a function of turbidity (liquid culture) or microbial activity parameters such as the rate of CO_2 evolution

(R_{CO_2}) or O_2 consumption (R_{O_2}) (liquid or solid substrate) (Wasileski et al., 1978; McAneney et al., 1979, 1980; Brown, 1978; Wilson and Griffin, 1975a; Griffin, 1978) or, for mycelial organisms (solid substrates) by direct measurement of radial growth (Sommers et al., 1970; Griffin, 1978). Concurrent mass balance analysis allows calculation of $Y_s^{\mu max(sat)}$ and $q_s^{\mu max(sat)}$:

$$Y_s^{\mu max(sat)} = \frac{1}{A_s^{\mu max(sat)}} = \frac{1}{A_{a-s}^{\mu max(sat)} + A_{d-s}^{\mu max(sat)}} = \frac{\mu_{max}}{q_s^{\mu max(sat)}} = \frac{\Delta X}{\Delta S} \quad [19.1]$$

$$A_{d-s}^{\mu max(sat)} = \frac{\Delta S_d}{X_{max(s)} - X_o} = \frac{S_o - (\Delta S_a + \Delta S_r)}{X_{max(s)} - X_o} \quad [19.2]$$

where ΔS_d, ΔS_a, and ΔS_r are, respectively, the amounts of substrate dissimilated, assimilated and remaining unused at growth cessation, as determined, for example for C-bound electron mass balance analysis (Harris and Adams, 1979). Cells from the late log/early stationary inflection may be characterized in terms of v_s^{max} and $K_{v(s)}^{(sat)}$ by standard radiotracer uptake techniques (Matin and Veldkamp, 1978; Brown et al., 1978). For $v_s^{max(sat)} \gg j_s^{(sat)}$: $v_s^{max(sat)} \cong q_s^{\mu max(sat)}$. The continuous culture technique may be used to determine the effect of ψ stress on the uptake and efficiency determinants of R_x under R_{sf}-limiting growth conditions ($S_m < S < S_{min}^{\mu max}$). In steady state, $\overline{S}$ (the concentration of growth-limiting S in the reactor and effluent) and $\overline{X}$ (the concentration of cells in the reactor and effluent) are fixed as a function of R_{sf} as controlled by S_o (the influent S concentration) and D (the culture dilution rate) $= \mu$ (Uden, 1967; Pirt, 1975; Matin and Veldkamp, 1978; Tempest and Neijssel, 1978). Accordingly, D, $\overline{S}$ and $\overline{X}$ may be substituted for μ, S and X in the R_x equations. Additional relevant derivations from these equations are:

$$Y_s^{(lim)} = \frac{1}{A_s^{(lim)}} = \frac{D}{q_s^{(lim)}} = \frac{\overline{X}}{S_o - \overline{S}} \quad [20.1]$$

$$A_s^{(lim)} = A_s^{m(lim)} + A_s^{b(lim)} = \frac{q_s^{m(lim)}}{D} + A_s^{b(lim)} = q_s^{m(lim)} T_D + A_s^{b(lim)} \quad [20.2]$$

$$A_s^{b(lim)} = A_{a-s}^{(lim)} + A_{d-s}^{b(lim)} = \frac{1}{Y_s^{b(lim)}} \quad [20.3]$$

$$Y_s^{\mu max(*)} = \frac{1}{A_s^{\mu max(*)}} = \frac{1}{A_s^{m(*)} + A_s^{b(*)}} = \frac{1}{q_s^{m(*)}/\mu_{max} + A_s^{b(*)}} \quad [21]$$

where $T_D = 1/D$ is the culture detention or replacement time. A plot of experimentally determined $A_s^{m(lim)}$ against T_D in the $0.05\ \mu_{max} < D < 0.9\ \mu_{max}$ range provides $q_s^{m(lim)}$ (slope) and $A_s^{b(lim)} = 1/Y_s^{b(lim)}$ (A_s intercept).

$Y_s^{\mu max(*)}$ is calculated using the batch culture-derived μ_{max} and assuming $q_s^{m(*)} = q_s^{m(lim)}$ and $A_s^{b(*)} = A_s^{b(lim)}$. Subsamples at specific D levels may be characterized for $v_s^{max(lim)}$ and $K_{s(v)}^{lim}$ by radiotracer techniques as for batch cultures. However, analytical problems of accurate $\overline{S}$ determination in the μM range, necessary for $K_{s(v)}$ determination, are severely compounded by the high stress solute concentrations in ψ stress systems. In addition, as for determination of the composition of ψ stressed cells, separation of cell biomass from the growth medium is a major experimental problem restricting attainment of valid yield measurements. Intracellularly-accumulated inducible compatible solutes and stress solutes are particularly sensitive to leakage loss during washing (Brown, 1978; Griffin, 1978). Use of isotonic wash solutions attenuates this problem (Brown, 1978), but complications with respect to wash solute carry-over contamination cannot be avoided and pose a particularly serious problem for dry weight-based growth yields (e.g. a −80 bar NaCl solution contains over 10% (W/V) salt; Table A-3a). The practical importance of intracellular solute leakage and wash solution carry-over depends on the magnitude and nature of the ψ stress, the mechanism used for intracellular ψ control and the biomass parameter used for yield characterization. For example, the dry weight composition (as an index of relative yield) of a −80 bar NaCl-stressed Class I organism would decrease from 127.4 to 103.5 g/mole cells if all the intracellularly-accumulated NaCl was leached out during washing, although cell organic and N content would be unchanged (Fig. 2; Table A-5). However, leakage of intracellularly-accumulated solutes from ψ-stressed Class II and IV organisms would cause substantial changes in cell organic and, for amino acid induction ψ responses, in cell N as well as in dry weight-based yields (Fig. 2; Table A-5). The cell separation problem may be circumvented by determination of cell yield in terms of cell N, cell C-bound e^-eq or cell organic C as a function of mass balance differences between total and filtrate samples of microbial cultures (Wasileski et al., 1978; Harris and Adams, 1979).

EXPERIMENTAL DATA

Growth rate reductions caused by inhibition of catabolic, and particularly anabolic, intermediary metabolism enzymes would be expected to be attenuated by the presence of key preformed metabolites, including compatible solutes and/or their immediate precursors, in the growth medium. The general observation that the more complex the growth medium, the greater the tolerance to ψ stress (Scott, 1957; Sommers et al., 1970; Griffin, 1977, 1978; Wasileski et al., 1978), supports the concept that the mode of action of ψ stress inhibition of microbial growth rates is exerted at the level of intracellular anabolic enzyme (q_{a-s}^{es}) inhibition. Further support is supplied by the increased growth rate tolerance to ψ stress shown by organisms such as *Salmonella oranienberg* in the presence of proline, a well-known compatible solute in its own right and also an immediate precursor of glutamate$^-$ (Christian and Hall, 1972), the compatible solute most widely used by procaryotes (Measures, 1975).

Quantitative information on the effect of ψ stress on the efficiency and rate determinants of assimilatory nutrient-limited growth is very limited. Under NaCl stress, Mg uptake is competitively inhibited (Tempest and Meers, 1968) but this is presumably more of a specific ion rather than a ψ effect. The relative importance of different nutrients (e.g. N, P, K vs. energy sources) as determinants of diffusion-limited growth in matric systems has yet to be critically defined (McAneney, 1979). However, microbial responses to specific nutrients provide qualitative information, and such data tend to support, for example, the commonly held concept that microorganisms are characteristically energy-limited in soil environments (Alexander, 1977).

Comprehensive data on the effect of ψ on the energy substrate efficiency and rate parameters controlling eucaryotic microbial growth are confined to the NaCl relations of *Saccharomyces cerevisiae* (Watson, 1970). A combination of aerobic batch and continuous culture procedures was used with glucose as the energy substrate and dry weight as the yield parameter (Watson, 1970). Batch culture μ_{max} and $Y_s^{\mu max(sat)}$ declined progressively over the ψ range evaluated (down to -70 bars), but $q_{gluc}^{\mu max(sat)}$ and thus presumably $v_{gluc}^{max(sat)}$, were not responsible for the μ_{max} reduction since $q_{gluc}^{\mu max(sat)}$ remained unchanged independent of ψ. A change in both $q_{gluc}^{m(lim)}$ (10-fold increase) and $Y_{gluc}^{b(lim)}$ (one and a half-fold decrease) occurred in continuous culture under -45 bar NaCl stress, indicative of a decrease in energy coupling efficiency ($\xi_{ATP/gluc}$) as well as an increase in q_{ATP}^{m}. Subsequent correction for the decrease in energy coupling efficiency, based on catabolic pathway considerations, showed a four-fold increase in q_{ATP}^{m} under ψ stress (Watson, 1970). The mechanistic response of *S. cerevisiae* to ψ stress has since (Brown, 1978) been established as involving intracellular accumulation of glycerol as a compatible solute; but the resulting change in cell composition (Table A-5) would exert only a relatively small effect on Y_{gluc}^{b}. Mechanistically, the decrease in μ_{max} under ψ stress was thus largely attributable to the increased q_{ATP}^{m} for osmoregulatory control, magnified by the decreased energy coupling efficiency, and to a lesser extent by the decreased Y_{gluc}^{b} (caused by the decreased $\xi_{ATP/gluc}$). Competitive growth of *S. cerevisiae* under subsaturating S would thus be severely reduced under saturating and subsaturating S conditions and the S_m would be increased substantially in accordance with the 10-fold increase in q_{gluc}^{m} (Eq. 15.2).

The solute ψ (ψ_s) relations of *Arthrobacter crystallopoietes*, representing an autochthonous Gr^+ soil bacterium well-adapted ecologically to fluctuating soil ψ and energy-limiting conditions, were characterized using a combination of batch and continuous culture approaches (Wasileski et al., 1978; Unluturk, 1977). The effect of NaCl-, KCl-, and Na_2SO_4- controlled ψ_s on the growth rate (turbidity-based μ_{max}) and growth yield $Y_s^{\mu max(sat)}$ (based on cell C-bound e^-, cell N and, for the non ψ-stressed control, cell C and dry weight) of ultimately succinate-limited (100 me^- eq succ$\cdot l^{-1}$ initial) batch cultures was determined. Growth rate declined from 0.12 to 0.02 hr^{-1} μ_{max} and growth yield from 7.3 to 5.8 e^-eq cells$\cdot$mole succ$\cdot l^{-1}$, from 0 to -80 bar NaCl- and KCl-controlled and -35 bar Na_2SO_4 ψ_s (extinction at -90 bar NaCl and KCl and -40 bar

Na_2SO_4) (Wasileski et al., 1978). Acclimatization of inocula to ψ stress did not affect subsequent growth response to ψ stress. The maintenance coefficient of 0.8 mmole suc•e^-eq $cells^{-1}$•hr^{-1} and the biosynthesis requirement (A^b_{succ}) of 130 mmoles suc•e^-eq $cells^{-1}$ shown by the unstressed control were relatively unaffected by ψ_s stress. Furthermore, the μ_{max} reductions in batch cultures under ψ_s stress were explainable solely on the basis of the increased maintenance energy arising from the associated μ_{max} reduction. Amino acid analysis of unstressed vs. ψ_s-stressed cells showed comparable high levels of glutamate (Wasileski, 1980). The response of *A. crystallopoietes* to cation concentration rather than to inorganic solute ψ_s per se, together with the unchanged growth efficiency and amino acid composition under ψ_s stress are consistent with an energy conserving mechanism involving passive equilibration of the ψ stress solutes across the cell membrane, characteristic of an osmoregulation Class III organism. Implicit in this mechanism, taking into consideration the relatively high growth tolerance of *A. crystallopoietes* to ψ_s, is that the intracellular enzymes of this organism show a relatively high tolerance to high concentrations of stress solutes, but this has yet to be tested directly.

The solute diffusion-independent, matric water potential (ψ_m) relations of *A. crystallopoietes* were evaluated as a function of the effect of polyethylene glycol (PEG)-controlled ψ_m on growth rate (μ_{max} based on turbidity and CO_2 evolution) and growth yield ($Y^{\mu max(sat)}_{succ}$ based on cell N) (McAneney et al., 1980). Neither μ_{max} nor growth yield was affected deleteriously over the 0 to −15 bar PEG-controlled ψ_m range evaluated, indicating successful cell wall ψ adjustment under ψ_m stress.

The ψ-independent, solute diffusion relations of *A. crystallopoietes* were evaluated as a function of growth rate (R_{CO_2}) response in a 0.42 g H_2O•g $soil^{-1}\Theta_w$/−1.0 bar ψ_m, fritted clay aggregate system amended with 1.3, 1.9, and 2.6 mmoles succinate C/100 g soil (McAneney et al., 1979; McAneney, 1979) where Θ_w is the soil water content. The background levels of organic carbon and N, although low, still proved too high to allow satisfactory mass balance analysis for yield determination (McAneney, unpublished). The nature of the dependence of R_{CO_2} on initial succinate concentration was consistent with diffusion-limited growth and had the same mathematical form as the solution to the problem of solute diffusion out of a porous sphere.

The effect of combined soil-controlled ψ_m and restricted solute diffusion on *A. crystallopoietes* was evaluated in −10 bar/0.25 g H_2O•g^{-1} and −15 bar/0.2 g H_2O•g^{-1} fritted clay systems in conjunction with the −1 bar/0.42 g H_2O•g^{-1} experiment (McAneney et al., 1979; McAneney, 1979). R_{CO_2} declined with decreasing ψ_m/Θ_w, from <2 days for 90% CO_2 evolution at −1 bar/0.42 g H_2O•g^{-1} to about 4 days at −15 bar/0.2 g H_2O•g^{-1}. This growth rate is much lower than would be predicted from the PEG-ψ_m relation of *A. crystallopoietes* [i.e., growth unaffected in the 0 to −15 bar ψ_m range (McAneney et al., 1980)], and in addition showed kinetics consistent with restricted solute diffusion. Accordingly, the predominant response of the arthrobacter to decreasing soil water in the −1 to −15 bar ψ_m range appeared to be a function of the effect of restricted solute diffusion caused by the associated reduction in Θ_w from 0.42 to 0.2 g H_2O/g soil.

The ψ_s relations of *Escherichia coli*, representing a Gr^- faecal indicator bacterium adapted for life in a relatively nutrient rich and unperturbed ψ environment, were characterized using an approach analogous to that used for the arthrobacter. The effect of NaCl-, KCl-, and Na_2SO_4-controlled ψ_s on the growth rate (μ_{max}) and growth yield ($Y_s^{\mu max(sat)}$) of ultimately succinate-, glucose-, yeast extract- or nutrient broth-limited (100 me^-eq substrate$\cdot l^{-1}$ initial) batch cultures was determined (Wasileski, 1980; Wasileski et al., 1978). Similar trends were shown for all salt-controlled ψ_s and C/energy source combinations. The higher unstressed μ_{max} shown on the richer media carried over to extend the ψ_s range for growth: growth extinction occurred at -40 to -50 bars on succinate and -60 to -80 bars on nutrient broth. On succinate, the μ_{max} declined from 0.32 hr^{-1} to 0.06 hr^{-1} and the growth yield from 7.2 to 5.4 e^-eq cells$\cdot$mole succ$\cdot^{-1}$, under -30 bar NaCl ψ_s stress. As for *A. crystallopoietes*, acclimatization of inocula to ψ stress did not affect subsequent growth responses to ψ stress (Wasileski, 1980). The maintenance coefficient (q_{succ}^m) increased from 0.55 to 1.00 mmole succ$\cdot e^-eq$ cells$^{-1}\cdot hr^{-1}$ and the biosynthesis requirement (A_{succ}^b) increased from 136 to 170 mmoles succ$\cdot e^-eq$ cells^{-1}, indicating both a decrease in energy coupling efficiency and an increase in osmoregulatory ATP maintenance energy under ψ stress. Maintenance coefficient correction for energy uncoupling showed that the osmoregulatory energy cost for increased solute gradient control under -30 bar NaCl ψ stress was relatively minor, increasing only from about 8 to 12 mmoles ATP$\cdot e^-eq$ cells$^{-1}\cdot hr^{-1}$ under -30 bar ψ stress, and essentially no maintenance energy increase was shown under -20 bar NaCl stress (Wasileski, 1980). A similar, essentially negligible, effect of 2% NaCl (about -15 bars) on maintenance energy (dry weight-based) was found by Neijsell and Tempest (1976) for *Klebsiella aerogenes* (osmoregulatory Class II, Table 3) and in effect by Aiking et al. (1977) for *Candida utilis* (osmoregulatory Class II?). Amino acid analysis of unstressed *E. coli* cells showed very low glutamate levels (relative to *A. crystallopoietes*); under -30 bar NaCl ψ_s stress the glutamate pool increased about 25 fold (Wasileski, 1980). The response of *E. coli* to ψ_s per se rather than to cation concentration, together with the increased glutamate pool and the increased energy requirement under ψ stress, is consistent with an energy-requiring osmoregulatory response involving intracellular accumulation of glutamate as an inducible compatible solute, and identifies *E. coli* as an osmoregulatory Class II organism.

The solute diffusion-independent, ψ_m relations of *E. coli* were evaluated as a function of the effect of PEG-controlled ψ_m on μ_{max} and growth yield (cell N-based) as for *A. crystallopoietes* (McAneney et al., 1980; McAneney, 1979). Unlike the arthrobacter, *E. coli* was highly sensitive to PEG-ψ_m: μ_{max} decreased 70% from 0 to -7.5 bars with extinction at -8 bars; the associated substantial growth yield reduction was not explainable on the basis of increased maintenance energy, but was more consistent with extensive cell death and lysis, indicative of an inability to regulate cell wall ψ under ψ_m stress. However, the precise mechanistic response of *E. coli* to PEG-ψ_m stress has yet to be defined unequivocally. The ψ-independent, solute diffusion relations of *E. coli* were evaluated as for *A. crystallopoietes* using a -1 bar fritted clay growth medium but amended with only one level of succinate (McAneney et al., 1979; McAneney, 1979). A similar non-exponential R_{CO_2} (rate of CO_2 evolution) was shown by both organisms, supporting the premise that restricted solute diffusion (rather than the μ_{max} of the individual organisms) was controlling growth rate in the −1 bar fritted clay system. The effect of combined soil-controlled ψ_m and restricted solute diffusion on *E. coli* was evaluated as for *A. crystallopoietes* in −5 bar/0.3 g $H_2O \cdot g^{-1}$, −7 bar/0.27 g $H_2O \cdot g^{-1}$ and −10 bar/0.25 g $H_2O \cdot g^{-1}$ fritted clay systems amended with succinate (McAneney et al., 1979; McAneney, 1979). The rate of CO_2 evolution declined rapidly with decreasing ψ_m from −5 to −7 bars and ceased at −10 bars. Taking into consideration the relatively rapid growth of *A. crystallopoietes* in the −10 bar/0.25 g $H_2O \cdot g^{-1}$ system and the severe effect of PEG-controlled ψ_m on *E. coli* (growth extinction at −8 bars), the response of *E. coli* to decreasing soil water in the −1 bar/0.42 g $H_2O \cdot g^{-1}$ to −10 bar/0.25 g $H_2O \cdot g^{-1}$ range appeared to be a combined function of ψ_m and restricted solute diffusion phenomena, with the former becoming increasingly more important with decreasing soil ψ_m/Θ_w such that growth extinction (possibly related to cell wall ψ adjustment problems) occurred at −10 bar soil ψ_m.

A basic ecological question at this stage concerns the practical implications of the differences in soil ψ relations between *A. crystallopoietes* and *E. coli*: Are these differences a reflection of soil vs. faecal, Gr^+ vs. Gr^-, Class II vs. Class III criteria, or are they related merely to unique physiological differences between the two organisms? From an experimental option standpoint, is growth response to NaCl vs. Na_2SO_4 a useful approach for characterizing the degree of energy-demanding osmoregulation under ψ stress? Similarly, is growth response to PEG a reliable index of cell wall ψ control mechanism under matric ψ stress?

Effect of Water Potential on Microbial Survival

Under conditions where the rate of substrate supply cannot meet the substrate requirement for growth ($R_{sf} = R_{sq} \leq R^m_{sq}$; $S \leq S_m$; $R_x \leq 0$), the microbial population becomes energy-starved, and survival becomes the primary physiological concern. Since all the exogenous substrate must now be used for maintenance of existing cells (so that none is left over for biosynthesis of new cells), the Y and A forms of the R_x equation are inappropriate for evaluating the biomass response of such non-growing cells to ψ stress.

BASIC EQUATIONS DEFINING THE DETERMINANTS OF MICROBIAL SURVIVAL

Physiologically-based relationships defining microbial survival and death rates are currently not available, although certain general trends are evident. Death rate is commonly very low relative to growth rate under optimum, saturating energy substrate growth conditions, but increases as μ declines with subsaturating S approaching S_m, the S level below which q_s is no longer adequate to support growth metabolism. As S declines below S_m, the microbial cells can either maintain growth transiently on endogenous reserves until death occurs because endogenous metabolism cannot meet the maintenance energy requirement for growth, or the cells can shift from growth metabolism (supported by $q_s^m = q_s^{m-v} + q_s^{m-b}$) to a less energy-demanding survival metabolism ($q_s^m = q_s^{m-v}$). The shift between growth and survival metabolism undergone by starving cells is frequently traumatic, especially for slow growing cells characteristically low in endogenous reserves, and gives rise to a higher mortality rate during the initial starvation period, commonly followed by a much lower death rate for cells successful in making the transition from growth to survival metabolism (Postgate and Hunter, 1962; Dawes, 1976).

The R_x and X equations for the case where the rate of substrate supply is exactly poised against the rate of substrate consumption for maintenance ($R_{sf} = R_{sq}^m$; $S = S_m$), thereby resulting in steady state maintenance of a specific biomass level, are as follows:

$$R_x = R_{sf} - R_{sq}^m = R_{sf} - q_s^m X_{max} = 0 \qquad [22]$$

$$X_{max}^{gr} = \frac{R_{sf}}{q_s^m} = \frac{R_{sf}\xi_{ATP/s}Ec}{q_{ATP}^m} = \frac{R_{sf}\xi_{ATP/s}Ec}{q_{ATP}^{m-v} + q_{ATP}^{m-b}} \qquad [23.1]$$

$$X_{max}^{surv} = \frac{R_{sf}}{q_s^{m-v}} = \frac{R_{sf}\xi_{ATP/s}Ec}{q_{ATP}^{m-v}} \qquad [23.2]$$

where X_{max}^{gr} and X_{max}^{surv} are the maximum biomass of cells that can be supported in a growth vs. a survival state of metabolism, respectively, for a given R_{sf}.

For the case where the rate of substrate supply is inadequate to meet the maintenance energy needs of the microbial population ($R_{sf} = R_{sq} < R_{sq}^m$; $S < S_m$), the maintenance energy requirement must be met by endogenous metabolism (cell decay), as attenuated by R_{sf}. The appropriate R_x and X equations are:

$$R_x = \frac{R_{sq} - R_{sq}^{m-v}}{\xi_{s/x}} = \frac{R_{sq}}{\xi_{s/x}} - R_x^{m-v} \qquad [24.1]$$

$$= -\frac{dX^{surv}}{dt} = -\lambda X^{surv} = \left(\frac{q_{d-s}}{\xi_{s/x}} - q_x^{m-v}\right) X^{surv}$$

$$= \left(\frac{q_{d-s}\xi_{ATP/s}}{\xi_{ATP/x}} - \frac{q_{ATP}^{m-v}}{\xi_{ATP/x}Ec}\right) X^{surv} \qquad [24.2]$$

$$= (v_s^{(surv)} - j_s^{(surv)} - q_x^{m-v})\, X^{surv}$$

$$= \left(v_s^{max(surv)}\frac{S}{S + K_{s(v)}^{(surv)}} - j_s^{(surv)} - \frac{q_{ATP}^{m-v}}{\xi_{ATP/x}Ec}\right) X^{surv} \qquad [24.3]$$

$$= \frac{R_{sf}}{\xi_{s/x}} - q^{m-v}X^{surv} = \frac{R_{sf}\xi_{ATP/s}}{\xi_{ATP/x}} - \frac{q_{ATP}^{m-v}X^{surv}}{\xi_{ATP/x}} \qquad [24.4]$$

$$X_t^{surv} = X_o^{surv} \exp(-\lambda t) \qquad [25]$$

$$t_{0.5} = \frac{0.69}{\lambda} \qquad [26]$$

where R_x^{m-v} is the rate of endogenous metabolism, expressed in biomass decay rate form, required to maintain the viability of X^{surv} biomass of cells; $\xi_{s/x}$ is the relative dissimilatory energy worth of the energy substrate compared to the dissimilatory energy worth of unit biomass of cell material; λ is the decay rate constant for the cell biomass resulting from endogenous metabolism; $t_{0.5}$ is the corresponding half life; $\xi_{ATP/x}$ is the amount of ATP generated per unit mass of cell material dissimilated; the "surv" superscript denotes survival metabolism properties. For $R_{sf} = 0$: $\lambda = q_x^{m-v}$ in the above equations.

Cell death rate (γ) is presumably related in some way to the decay rate, but precise relationships are not currently available. Using F as an ill-defined proportionality factor relating λ to γ ($\gamma = F\lambda$):

$$R_{Nxv} = -\frac{dN_{xv}}{dt} = -\gamma N_{xv} = -F\lambda N_{xv} \qquad [27]$$

where N_{xv} is the number of viable cells.

The effect of ψ stress on microbial survival is a function of the effect of the stress on the efficiency ($\xi_{ATP/s}$, $\xi_{ATP/x}$, Ec, q_{ATP}^{m-v}, q_{ATP}^{m-b}) and uptake rate (v_s^{max}, j_s, $K_{s(v)}$) determinants of decay (λ) and death (γ) rate. Restricted information on the effect of ψ stress on survival of starved Gr^- procaryotes (Postgate and Hunter, 1962; Wasileski and Harris, 1979) indicates accelerated death under solute-controlled ψ stress. The increased osmoregulatory energy requirement for growth of *E. coli* under ψ_s stress apparently carried over to its endogenous metabolism, as indicated by accelerated decay and death under ψ stress (Wasileski and Harris, 1979), although ψ stress interference with the shift from growth to survival metabolism may also have been involved. In contrast, the death rate of *A. crystallopoietes*, representing a relatively salt tolerant, osmoregulatory Class III organism (no increase in maintenance coefficient under ψ stress), was not accelerated under -40 bar NaCl ψ stress (Wasileski, 1980). Direct information on the effect of matric ψ stress on the survival of starved microbial cells in the ψ range supporting microbial growth is not available. Under extreme matric ψ stress dessication occurs. Microbial survival under dessication stress has been widely studied, although relationships between death rate and dessication conditions are currently ill-defined. Theoretically, accumulation of intracellular ψ-controlling solutes offers a protective effect against severe plasmolytic dehydration for downshocked cells (Fig. 2). Relatedly, microbial cells grown under solute ψ stress, and thereby containing correspondingly high intracellular concentrations, frequently show lower death rates following subsequent dessication, than comparable non ψ-stressed cells (Chen and Alexander, 1973). Comparative studies of the dessication-survival potential of organisms representing different osmoregulatory classes are not available.

Effect of Water Potential on Microbial Activity

This section evaluates the principles controlling the effect of ψ stress on microbial activity. The water potential relations of specific microbially-mediated transformations are reviewed elsewhere (Sommers et al., Chapter 3). Whether ψ stress decreases or increases, microbial activity depends on the relative importance of the ψ stress effects on the different physiological determinants of activity:

For $R_x > 0$:

$$R_{zq} = R_x A_z = R_x(A_{d-z} - A_{a-z}) = R_x(A_{d-s}\alpha_{z/s}) - R_x A_{a-z}$$

$$= q_z X = [(q_s^m + \mu A_{d-s}^b)\alpha_{z/s} - \mu A_{a-z}]X$$

$$= \left[\left(\frac{q_{ATP}^m + \mu A_{ATP}^b}{\xi_{ATP/s}Ec}\right)\alpha_{z/s} - \mu(A_{a-z}^{es} + A_{a-z}^{st})\right] X \qquad [28]$$

$$Z_t = (X_t - X_o)A_z$$

$$= (X_o\exp(\mu t) - X_o)\left[\left(\frac{q_s^m}{\mu} + A_{d-s}^b\right)\alpha_{z/s} - A_{a-z}\right] \qquad [29.1]$$

$$Z_{max(s)} = (X_{max(s)} - X_o)\left[\left(\frac{q_s^m}{\mu} + A_{d-s}^b\right)\alpha_{z/s} - A_{a-z}\right]$$

$$= \left[\frac{S_o}{(q_s^m/\mu) + A_{d-s}^b + A_{a-s}}\right]\left[\left(\frac{q_s^m}{\mu} + A_{d-s}^b\right)\alpha_{z/s} - A_{a-z}\right]$$

$$\rightarrow S_o\alpha_{z/s} \text{ (as } \mu \rightarrow 0) \qquad [29.2]$$

For $R_x = 0$

$$R_{zq} = R_{sf}\alpha_{z/s} = R_{sq}^m\alpha_{z/s} = q_z^m X = q_s^m\alpha_{z/s}X = \frac{q_{ATP}^m}{\xi_{ATP/s}Ec}\alpha_{z/s}X \qquad [30]$$

$$Z_t = R_{zq}t = q_s^m\alpha_{z/s}Xt = \frac{q_{ATP}^m}{\xi_{ATP/s}Ec}\alpha_{z/s}Xt \qquad [31]$$

For $R_x < 0$:

$$R_{zq} = R_{sf}\alpha_{z/s} - \left(\frac{R_{sf}}{\xi_{s/x}} - R_x^m\right)A_{a-z} = R_{d-sq}\alpha_{z/s} - \lambda A_{a-z}X$$

$$= q_z X = \left[q_{d\text{-}s}\alpha_{z/s} - \left(\frac{q_{d\text{-}s}}{\xi_{s/x}} - q_x^m \right) A_{a\text{-}z} \right] X$$

$$= \left[q_{d\text{-}s}\alpha_{z/s} - \left(\frac{q_{d\text{-}s}\xi_{ATP/s}}{\xi_{ATP/x}} - \frac{q_{ATP}^m}{\xi_{ATP/x}Ec} \right) (A_{a\text{-}z}^{es} + A_{a\text{-}z}^{st}) \right] X \qquad [32]$$

$$Zt = R_{sf}\alpha_{z/s}t + (X_o - X_t)A_{a\text{-}z}$$

$$= R_{sf}\alpha_{z/s}t + [X_o - X_o\exp(-\lambda t)](A_{a\text{-}z}^{es} + A_{a\text{-}z}^{st}) \qquad [33]$$

$R_{zq} = (dZ/dt)_q$ is the rate of change of Z, the concentration of a biochemical reactant (other than the electron donor/organic C source, S) or product, associated with microbial growth and/or survival; q_z is the net specific rate of Z consumption or production; q_z^m is the specific rate of Z production associated with maintenance; A_z is the proportionality of Z with respect to microbial biomass (X); $A_{d\text{-}z}$ and $A_{a\text{-}z}$ are the dissimilatory and assimilatory proportionalities of Z to biomass; $A_{a\text{-}z}^{es}$ is the proportionality of Z to the essential cell material component of microbial biomass and $A_{a\text{-}z}^{st}$ is the analogous proportionality of Z to the storage and/or ψ stress-related cell material component; $\alpha_{z/s}$ is the proportionality of Z to the electron donor/organic C substrate (S); Z_t and $Z_{max(s)}$ are the net amounts of Z consumed or produced after time t and from a given amount of S, respectively.

For simple assimilatory nutrient transformations, such as N immobilization during microbial growth with organic C as the sole electron donor/C source, equations 28 and 29 become:

$$R_{NH_4^+} = -q_{NH_4^+}X = -\mu A_{a\text{-}NH_4^+}X = -\mu(A_{a\text{-}NH_4^+}^{es} + A_{a\text{-}NH_4^+}^{st})X \qquad [34]$$

$$[NH_4^+]_{max(s)} = (X_{max(s)} - X_o)A_{a\text{-}NH_4^+}$$

$$= S_oY_sA_{a\text{-}NH_4^+} = \frac{S_oA_{a\text{-}NH_4^+}}{(q_s^m/\mu) + A_{d\text{-}s}^b + A_{a\text{-}s}}$$

$$= \frac{S_o(A_{a\text{-}NH_4^+}^{es} + A_{a\text{-}NH_4^+}^{st})}{\dfrac{q_{ATP}^m}{\mu\xi_{ATP/s}Ec} + \dfrac{A_{ATP}^b}{\xi_{ATP/s}Ec} + A_{a\text{-}s}^{es} + A_{a\text{-}s}^{st}} \qquad [35]$$

where $R_{NH_4^+}$ is the net rate of change of NH_4^+; $q_{NH_4^+}$ is the specific rate of NH_4^+ assimilation; $A_{a\text{-}NH_4^+}$ is the assimilatory NH_4^+ requirement (N content); $A_{a\text{-}NH_4^+}^{es}$ and $A_{a\text{-}NH_4^+}^{st}$ are the assimilatory NH_4^+ requirements for biosynthesis of essential cell materials and biosynthesis of storage and/or ψ stress-related materials, respectively; $[NH_4^+]_{max(s)}$ is the amount of NH_4^+ assimilated for a given amount of energy/C substrate. Cell composition is inevitably affected by ψ stress (Fig. 2; Table A-5). However, although induction of amino acids by Class II and IV organisms under ψ stress will cause an increase in $A_{a\text{-}NH_4^+}$ thereby tending to increase $R_{NH_4^+}$ (Eq. 34), the associated decline in growth yield will tend to cause a decrease in the total amount of NH_4^+ immobilized (Eq. 35). Furthermore, any ψ stress inhibition of μ (by whatever mechanism) will cause a decrease in growth efficiency and thus also in X, thereby resulting in a further tendency towards reduction in the rate of NH_4^+ immobilization. In practice, ψ stress (particularly ψ_m stress because of the restricted diffusion-S limitation phenomenon) will likely exert a greater effect on μ than on A_{a-z}, so that the overall effect of ψ stress will tend to be a reduction in the rate and extent of nutrient immobilization.

With organic N as the electron donor/C source, NH_4^+ assimilation during microbial growth is counterbalanced by generation of NH_4^+ as a by-product of organic N dissimilation, and Eq. 28 and 29 become:

$$\begin{aligned} R_{NH_4^+} &= R_x(A_{d\text{-}s}\alpha_{NH_4^+/s}) - R_x A_{a\text{-}NH_4^+} \\ &= q_{NH_4^+}X = [(q_s^m + \mu A_{d\text{-}s}^b)\alpha_{NH_4^+/s} - \mu A_{a\text{-}NH_4^+}]X \\ &= \left[\frac{(q_{ATP}^m + \mu A_{ATP}^b)\alpha_{NH_4^+/s}}{\xi_{ATP/s}Ec} - \mu(A_{a\text{-}NH_4^+}^{es} + A_{a\text{-}NH_4^+}^{st}) \right] X \end{aligned} \qquad [36]$$

$$\begin{aligned} [NH_4^+]_{max(s)} &= \left[\frac{S_o}{(q_s^m/\mu) + A_{d\text{-}s}^b + A_{a\text{-}s}} \right] \\ &\quad \left[\left(\frac{q_s^m}{\mu} + A_{d\text{-}s}^b \right) \alpha_{NH_4^+/s} - (A_{a\text{-}NH_4^+}^{es} + A_{a\text{-}NH_4^+}^{st}) \right] X \\ &\rightarrow S_o \alpha_{NH_4^+/s} \qquad (\text{as } \mu \rightarrow 0) \end{aligned} \qquad [37]$$

where $R_{NH_4^+}$ is the net rate of change of NH_4^+; $\alpha_{NH_4^+/s}$ is the amino N (−3 valent N) content of the organic N substrate; $q_{NH_4^+}$ is the net specific rate of NH_4^+ immobilization or generation; $[NH_4^+]_{max(s)}$ is the amount of NH_4^+ immobilized or generated for a given amount of organic N substrate. An increase in osmoregulatory energy for solute gradient control under ψ stress will tend to increase the maintenance energy coefficient and thereby shift the $R_{NH_4^+}$ balance in favor of organic N mineralization (Eq. 36). Similarly, a ψ stress-caused reduction in energy efficiency (decreased $\xi_{ATP/s}Ec$) will favor net NH_4^+ generation (Eq. 36 and 37). A decrease in μ caused by inhibition of the rate and/or reduction of the efficiency determinants under ψ stress will decrease both the rate of NH_4^+ assimilation and generation (Eq. 36), but will favor net mineralization until, as $\mu \rightarrow 0$ under ψ stress, the balance between immobilization and mineralization shifts entirely towards mineralization in line with the increasing importance of the maintenance energy component with decreasing μ (Eq. 37).

For simple cell decay-related nutrient transformations, such as N mineralization during cell decay caused by exhaustion of the energy source, Eq. 32 and 33 become:

$$R_{NH_4^+} = q_{NH_4^+}X = -\lambda A_{a-N}X = -q_x^{m-v}A_{a-N}X$$

$$= \frac{q_{ATP}^{m-v}}{\xi_{ATP/x}Ec}(A_{a-N}^{es} + A_{a-N}^{st})X \qquad [38]$$

$$[NH_4^+]_t = (X_o - X_t)A_{a-N} = [X_o - X_o\exp(-\lambda t)]A_{a-N}$$

$$= \left[X_o - X_o\exp\left(-\frac{q_{ATP}^{m-v}\,t}{\xi_{ATP/s}Ec}\right)\right](A_{a-N}^{es} + A_{a-N}^{st}) \qquad [39]$$

where $q_{NH_4^+}$ is the specific rate of ammonification during cell decay; A_{a-N} is the cell N content; A_{a-N}^{es} and A_{a-N}^{st} are the cell N fractions associated with essential cell materials and storage and/or ψ stress-related materials, respectively. A ψ stress response involving a requirement for additional osmoregulatory energy (increase in q_{ATP}^{m-v}) and/or interference with the efficiency of energy generation and use (decreased $\xi_{ATP/s}Ec$) will result in an increase in the decay rate (λ) and thus will increase the rate and extent of cell organic N mineralization (Eq. 38 and 39). Equations 28 to 33 apply to diverse aerobic and anaerobic, heterotrophic and autotrophic transformations involving the dissimilatory oxidation of electron donors and release of oxidized products; the dissimilatory reduction of electron acceptors and release of reduced products; and transformations reflecting combined assimilatory and dissimilatory reactions (e.g. organic N use as an electron donor for energy and as an assimilatory source of N, reflecting competing N mineralization and N immobilization reactions). The effect of ψ stress on the rate and extent of assimilatory, dissimilatory and combined assimilatory/dissimilatory reactions will depend on the relative importance of the ψ stress on the rate and efficiency determinants of biochemical activity (Eq. 28 to 33), but in general will likely result in a reduction in the rate of biochemical transformations and a shift towards net mineralization, because of the apparent general dominance of μ inhibition as the major physiological effect of both ψ per se and (restricted solute diffusion-dominated) ψ_m/Θ_w stress on microbial growth.

SUMMARY

The basic response of a microorganism to ψ stress is dominated by the biophysical need of the organism to attain ψ equilibrium with its environment. From a mechanistic standpoint, the total ψ of a microbial substrate is composed of: 1) a permeant solute-controlled component (commonly called the osmotic or solute ψ, ψ_s), which is mechanistically defined from a biological cell standpoint as the substrate ψ due to water sorption by dissolved solutes permeant (by passive and/or active transport) with respect to cell wall/membranes, and 2) a nonpermeant solute-controlled component (commonly called the matric ψ, ψ_m), which is the substrate ψ due to water sorption by dissolved solutes impermeant with respect to cell wall/membranes (impermeant solute-controlled ψ_m), and/or the substrate ψ due to water sorption by particulate surfaces (matrix-controlled ψ_m). The substrate ψ_s component is composed of ψ_s subcomponents arising from the different forms and amounts of solutes dissolved in the aqueous phase of the substrate. Environmentally, NaCl and sucrose are the most important stress solutes; other solutes useful for mechanistic comparison with NaCl and sucrose include KCl (K^+ vs. Na^+ specific ion effects), Na_2SO_4 (ψ vs. Na^+ specific ion effects) and glycerol (relatively highly permeant solute with respect to biological membranes). The substrate matrix-controlled ψ_m component arises from relatively short range adsorption forces between particulate surfaces and water molecules, and consequently becomes increasingly more important as the substrate water content (Θ_w decreases. However, since Θ_w dictates not only matrix-controlled ψ_m but also the rate of solute diffusion through the aqueous phase of the substrate, a change in matrix-controlled ψ_m is inevitably accompanied by a change in the solute diffusion properties of the substrate. The substrate impermeant solute-controlled ψ_m component is of ill-defined, but probably minor, importance in most microbial substrates. However, experimental use of impermeant solutes (e.g., polyethylene glycol) allows characterization of the ψ_m per se relations of microorganisms, information needed for establishment of the relative importance of ψ_m per se vs. solute diffusion as primary determinants of the effect of substrate water on microbial growth and activity in solid substrates such as soil.

From a ψ standpoint a microbial cell consists structurally of a protoplast (bounded by a semipermeable membrane), and a retaining cell wall (of minimal selective solute retention properties). The total ψ of a microbial protoplast is composed of a dissolved solute (ψ_s), bound water (ψ_m) and turgor pressure [$\psi_p = \psi_t - (\psi_s + \psi_m)$] component. The ψ_s component is further subdivided according to the forms and amounts of: dissolved basal metabolites; substrate-derived stress solutes; inducible compatible solutes (solutes less inhibitory to metabolic processes than stress solutes such as NaCl, that are accumulated intracellularly as an induced counter-balancing response to the extracellular presence of stress solutes); and constitutive compatible solutes (compatible solutes constitutively present in the cell in relatively high concentration independent of the presence of extracellular stress solutes, and thereby necessitating the existence of a strong cell wall to withstand the inevitably high resultant ψ_p). Acidic amino acids such as glutamate (boundary level of about −20 bars), with K^+ as the counterbalancing cation, are favored by procaryotes as inducible (e.g. certain Gr^- bacteria) and constitutive (e.g. Gr^+ bacteria) compatible solutes, with neutral amino acids such as proline and gamma amino butyrate (about −20 bar max) being used as supplementary inducible compatible solutes. Eucaryotes also use amino acids, but tend to prefer polyols such as arabitol (about −20 bar max) as a constitutive compatible solute (e.g. thick-walled yeasts) and glycerol (about −60 bar max) as an inducible compatible solute.

The relative importance of the cell wall ψ components involved in satisfying the ψ mass balance requirement that the total cell wall ψ be equal to the total ψ of the enclosed protoplast and the total ψ of the surrounding environment is currently ill-defined.

Microorganisms may be classified according to their osmoregulatory strategy for growth and survival under ψ stress. Osmoregulatory Class I microorganisms are potentially the most xerosensitive, particularly with respect to ψ_m perturbations, since they cannot induce compatible solutes under ψ stress nor do they accumulate compatible solutes constitutively. However, relatively salt-tolerant organisms cannot be excluded from this class since the intracellular enzymes of many organisms are not highly salt-sensitive. At the other end of the physiological spectrum are osmoregulatory Class IV organisms. By virtue of their capacity for constitutive production of compatible solutes (with the associated property of a strong cell wall/membrane complex) combined with their capacity for supplemental induction of additional compatible solutes, these organisms are capable of withstanding extreme downshock (plasmolytic) and upshock (plasmoptic) ψ stress, as well as being capable of sustained growth under low ψ conditions. Implicit in this classification approach is that the strategies used by microorganisms for growth and survival under ψ stress reflect the nature and extent of ψ perturbations of their natural habitats and, from a predictive standpoint, define the potential of microorganisms for growth and survival as a function of the specific ψ perturbation and solute properties of foreign habitats.

The growth, survival and activity response of a microorganism to ψ stress is dictated by the effect of the stress on the physiological and substrate supply determinants of the rate and efficiency of growth and maintenance. Whether a ψ stress causes a reduction in growth rate under saturating substrate conditions (μ_{max}) depends on whether the unstressed μ_{max} is limited by the rate of biosynthesis, the rate of energy generation or the rate of substrate uptake, as well as being a function of the nature, severity and mode of action of the ψ stress. For example, an organism of unstressed μ_{max} limited by the rate of biosynthesis, will be capable of absorbing (without decreasing its μ_{max}) a certain increase in osmoregulatory maintenance energy requirement for increased concentration gradient control under ψ stress, and a certain increase in the dissimilatory (increased energy) and assimilatory (inducible compatible solute accumulation) substrate requirement, as a function of the "buffer capacity" of its energy generation and substrate uptake rate properties. However, the efficiency of substrate use and thus population density will be reduced. Water stress effects on the rate and efficiency of substrate use become increasingly more important under substrate limiting conditions and may play a major role in determining competitive growth and survival in characteristically energy-limited substrates such as soil.

Conceptual approaches and experimental techniques are reasonably well defined for evaluating the ψ_s relations of microorganisms. However, most experimental data are limited to single parameter (usually growth rate) descriptions of the ψ_s relations of microorganisms; comprehensive data on the effect of ψ_s on the specific rate and efficiency determinants of microbial growth and survival are currently very limited. Virtually nothing is known about the ψ_m per se relations of microorganisms, although theoretical considerations and limited experimental data suggest that cell wall ψ adjustment phenomena may play a dominant role. Description of the water relations of microorganisms in solid substrates such as soil necessitates consideration of interrelated ψ_s, ψ_m per se and solute diffusion phenomena. For high water contents (where ψ_s and ψ_m are minimal), solute diffusion dominates the microbial response. However, current understanding of solute diffusion-microbial growth interactions, even for the restricted case of high ψ, is extremely inadequate, reflecting the lack of attention paid classically to this biophysical topic by microbiologists. With decreasing substrate water content, ψ_s and particularly ψ_m per se stress, as well as restricted solute transport, become increasingly more severe; consequently, establishment of the relative importance of solute transport as a determinant of the response of a specific microorganism to substrate water requires knowledge of the ψ_s and ψ_m per se relations of the organism.

Recognition of the mechanistic nature of the effect of soil water on microbial growth, survival and activity in soil, and related characterization of microbial physiological properties in relevant rate and efficiency terms, is essential for ultimate development of predictive models of ecological interactions between indigenous and invader soil microorganisms and quantitative description and prediction of microbially-mediated transformations under characteristically water-perturbed field soil conditions.

ACKNOWLEDGMENTS

Contribution from the College of Agricultural and Life Sciences, Univ. of Wisconsin, Madison, and supported in part by NSF DEB75-18582A01. The author is deeply indebted to W. R. Gardner for invaluable conceptual advice on the water potential balance and microbial kinetics aspects of the review, and to S. S. Adams for technical assistance.

LITERATURE CITED

1. Adebayo, A.A., and R. F. Harris. 1971. Fungal growth responses to osmotic as compared to matric water potential stress. Soil Sci. Soc. Am. Proc. 35:465–469.
2. ———, ———, and W. R. Gardner. 1971. Turgor pressure of fungal mycelia. Trans. Br. Mycol. Soc. 57:145–151.
3. Aiking, H., G. J. Van Holst, and D. W. Tempest. 1977. The influence of different carbon sources and medium osmolarity on the potassium requirements of *Candida utilis* NCYC 321, growing in continuous culture. Arch. Microbiol. 115:79–84.
4. Alexander, M. 1977. Introduction to soil microbiology. John Wiley and Sons, New York.
5. Baver, L. D., W. H. Gardner, and W. R. Gardner. 1972. Soil physics. John Wiley and Sons, New York.
6. Bovell, C. R., L. Packer, and R. Helgerson. 1963. Permeability of *Escherichia coli* to organic compounds and inorganic salts measured by light-scattering. Biochim. Biophys. Acta. 75:257–266.
7. Brown, A. D. 1976. Microbial water stress. Bacteriol. Rev. 40:803–846.

8. ————. 1978. Compatible solutes and extreme water stress in eucaryotic microorganisms. *In* A. H. Rose and J. G. Morris (ed.) Adv. Microbial Physiol. 17:181–242.

9. Brown, E. J., R. F. Harris, and J. F. Koonce. 1978. Kinetics of phosphate uptake by aquatic microorganisms: Deviations from a simple Michaelis-Menten equation. Limnol. Oceanogr. 23:26–34.

10. Buchanan, R. E., and N. E. Gibbons. 1975. Bergey's manual of determinative bacteriology. Eighth edition. The Williams and Wilkins Company, Baltimore, Md.

11. Chen, M., and M. Alexander. 1973. Survival of soil bacteria during prolonged dessication. Soil Biol. Biochem. 5:213–221.

12. Christian, J. H. B., and J. M. Hall. 1972. Water relations of *Salmonella oranienberg*: Accumulation of potassium and amino acids during respiration. J. Gen. Microbiol. 70: 497–506.

13. ————, and J. A. Waltho. 1964. The composition of *Staphylococcus aureus* in relation to the water activity of the growth medium. J. Gen. Microbiol. 35:204–213.

14. Dawes, E. A. 1976. Endogenous metabolism and the survival of starved procaryotes. Symp. Soc. Gen. Microbiol. 26:19–53.

15. Dubey, H. D. 1968. Effect of soil moisture on nitrification. Can. J. Microbiol. 14:1348–1350.

16. Dundas, I. E. D. 1977. Physiology of *Halobacteriaceae*. Adv. Microb. Physiol. 15:85–120.

17. Epstein, W., and S. G. Schultz. 1968. Ion transport and osmoregulation in bacteria. *In* L. B. Cruze (ed.) Microbiol protoplasts, spheroplasts, and L-forms. William and Wilkins Co., Baltimore, Md.

18. Farrar, J. F. 1976. The lichen as an ecosystem: observation and experiment. p. 385–406. *In* D. H. Brown, D. Hawksworth, and R. H. Bailey (ed.) Progress and problems in lichenology. Academic Press, London.

19. Gardner, W. R., and R. F. Harris. 1980. Kinetics of bacterial growth, survival, and activity in soil: II. Restricted solute availability. Am. Soc. Agron. Abstr.

20. Gradman, D. 1977. Potassium and turgor pressure in plants. J. Theor. Biol. 65:597–599.

21. Griffin, D. M. 1969. Soil water in the ecology of fungi. Ann. Rev. Phytopathol. 7:289–310.

22. ————. 1972. Ecology of soil fungi. Syracuse Univ. Press, Chapman and Hall, London.

23. ————. 1977. Water potential and weed-decay fungi. Ann. Rev. Phytopathol. 15: 319–329.

24. ————. 1978. Effect of soil moisture on survival and spread of pathogens. *In* T. T. Kozlowski (ed.) Water deficits and plant growth. Vol. V:175–197.

25. Harris, R. F., and S. Adams. 1979. Determination of the carbon-bound electron composition of microbial cells and metabolites by dichromate oxidation. Appl. Environ. Microbiol. 37:237–243.

26. ————, and W. R. Gardner. 1980. Kinetics of bacterial growth, survival and activity in soil: I. Unrestricted solute availability. Am. Soc. Agron. Abstr.

27. Hsiao, T. C., E. Acevedo, E. Fereres, and D. W. Henderson. 1976. Stress metabolism: Water stress, growth, and osmotic adjustment. Phil. Trans. R. Soc. Lond. B. 273:479–500.

28. Justice, J. K., and R. L. Smith. 1962. Nitrification of ammonium sulfate in a calcareous soil as influenced by combinations of moisture, temperature, and levels of added nitrogen. Soil Sci. Soc. Am. Proc. 26:246–250.

29. Keller, P. 1969. The effect of sodium chloride and sulphate on sulphur oxidation in soil. Plant Soil 30:15–21.

30. Knowles, C. J., and L. Smith. 1971. The relationship between substrate-induced respiration and swelling in *Azotobacter vinelandii*. Biochim. Biophys. Acta 234:153–161.

31. Koujima, I., H. Hayashi, K. Tomochika, A. Okabe, and Y. Kanemasa. 1978. Adaptational change in proline and water content of *Staphylococcus aureus* after alteration of environmental salt concentration. Appl. Environ. Microbiol. 35:467–470.

32. Kouyeas, V. 1964. An approach to the study of moisture relations of soil fungi. Plant Soil 20:351–363.

33. Lanyi, J. K. 1978. Light energy conversion in *Halobacterium halobium*. Microbiological Rev. 42:682–706.

34. Leistner, L., and W. Rodel. 1976. Inhibition of microorganisms in food by water activity. *In* F. A. Skinner and W. B. Hugo (ed.) Inhibition and inactivation of vegetative microbes. Academic Press, New York.

35. Makemson, J. C., and J. W. Hastings. 1979. Glutamate functions in osmoregulation in a marine bacterium. Appl. Environ. Microbiol. 38:178–180.

36. Marshall, B. J., D. F. Ohye, and J. H. B. Christian. 1971. Tolerance of bacteria to high concentration of NaCl and glycerol in the growth medium. Appl. Microbiol. 21:363–364.

37. Matin, A., and H. Veldkamp. 1978. Physiological basis of the selective advantage of a *Spirillum* sp. in a carbon-limited environment. J. Gen. Microbiol. 105:187–197.

38. McAneney, K. J. 1979. Bacterial growth in soils: Effects of matric water potential and nutrient transport. Ph.D. thesis, Univ. of Wisconsin, Madison.

39. ————, W. R. Gardner, and R. F. Harris. 1979. Matric water relations of bacteria. Am. Soc. Agron. Abstr.

40. ————, and R. F. Harris, and W. R. Gardner. 1980. Bacterial water relations using polyethylene glycol 4000. Soil Sci. Soc. Am. Submitted.

41. Measures, J. C. 1975. Role of amino acids in osmoregulation of non-halophilic bacteria. Nature 257:398–400.

42. Mexal, J., and C. P. P. Reid. 1973. The growth of selected mycorrhizal fungi in response to induced water stress. Can. J. Bot. 51:1579–1588.

43. Mitchell, P., and J. Moyle. 1956. Osmotic function and structure in bacteria. Soc. Gen. Microbiol. Symp. 6:150–180.

44. Moser, U. S., and R. V. Olsen. 1953. Sulfur oxidation in four soils as influenced by soil moisture tension and sulfur bacteria. Soil Sci. 76:251–257.

45. Neijssel, O. M., and D. W. Tempest. 1976. The role of energy-spilling reactions in the growth of *Klebsiella aerogenes* NCTC 418 in aerobic chemostat culture. Arch. Microbiol. 110:305–311.

46. Norrish, R. S. 1966. An equation for the activity coefficients and equilibrium relative humidities of water in confectionary syrups. J. Food Technol. 1:25–39.

47. Pirt, S. J. 1975. Principles of microbe and cell cultivation. John Wiley and Sons, New York.

48. Pitt, J. I. 1975. Xerophilic fungi and the spoilage of foods of plant origin. p. 273–307. *In* R. B. Duckworth (ed.) Water relations of foods. Academic Press, London.

49. ————, and A. D. Hocking. 1977. Influence of solute and hydrogen ion concentration on the water relations of some xerophilic fungi. J. Gen. Microbiol. 101:35–40.

50. Postgate, J. R., and J. R. Hunter. 1962. The survival of starved bacteria. J. Gen. Microbiol. 29:233–263.

51. Prior, B. A. 1978. The effect of water activity on the growth and respiration of *Pseudomonas fluorescens*. J. Appl. Bact. 44:97–106.

52. Reichman, G. A., D. L. Grunes, and F. G. Viets, Jr. 1966. Effect of soil moisture on ammonification and nitrification of two northern plains soils. Soil Sci. Soc. Am. Proc. 30:363–366.

53. Robinson, R. A., and R. H. Stokes. 1959. Electrolyte solutions. Academic Press Inc., New York.

54. Rose, A. H. 1976. Osmotic stress and microbial survival. Soc. Gen. Microbiol. Symp. 26:155–182.

55. Schobert, B. 1977. Is there an osmotic regulatory mechanism in algae and higher plants? J. Theor. Biol. 68:17–26.

56. Scott, W. J. 1957. Water relations of food spoilage microorganisms. Adv. Food Res. 7: 83–127.

57. Shilo, M. 1979. Life at low water activities. p. 15–135. *In* strategies of microbial life in extreme environments. Verlag Chemie, Weinheim, New York.

58. Sindhu, M. A., and A. H. Cornfield. 1967. Effect of sodium chloride and moisture content on ammonification and nitrification in incubated soil. J. Sci. Food Agric. 18:505–506.

59. Smith, D. C. 1979. Is a lichen a good model of biological interactions in nutrient-limited environments? p. 291–303. *In* M. Shilo (ed.) Strategies of microbial life in extreme environments. Verlag Chemie, Weinheim, New York.

60. Sommers, L. E., R. F. Harris, F. N. Dalton, and W. R. Gardner. 1970. Water relations of three root-infecting *Phytophthora* species. Phytopathol. 60:932–934.

61. Steinborn, J., and R. J. Roughley. 1975. Toxicity of sodium and chloride ions to *Rhizobium* spp. in broth and peat culture. J. Appl. Bact. 39:133–138.

62. Tempest, D. W., and J. L. Meers. 1968. The influence of NaCl concentration of the medium on the potassium content of *Aerobacter aerogenes* and on the interrelationships between potassium, magnesium, and ribonucleic acid in the growing bacteria. J. Gen. Microbiol. 54:319–325.

63. ————, ————, and C. M. Brown. 1970. Influence of environment on the content and composition of microbial free amino acid pools. J. Gen. Microbiol. 64:171–185.

64. ————, and O. M. Neijssel. 1978. Eco-physiological aspects of microbial growth in aerobic nutrient-limited environments. Adv. Microbiol. Ecol. 2:105–153.

65. Thauer, R. K., K. Jungerman, and K. Decker. 1977. Energy conservation in chemotrophic anaerobic bacteria. Bacteriol Rev. 41:100–180.

66. Uden, N. van. 1967. Transport-limited growth in the chemostat and its competitive inhibition; a theoretical treatment. Arch. Microbiol. 58:145–154.

67. Unluturk, A. 1977. Effect of osmotic water stress on the growth rate and growth yield of *Arthrobacter crystallopoietes*. Ph.D. thesis, Univ. of Wisconsin, Madison.

68. Veldkamp, H. 1976. Continuous culture in microbial physiology and ecology. Meadowfield Press, Durham, England.

69. Wasileski, J. M. 1980. Solute-controlled water potential relations of *Escherichia coli*. Ph.D. Thesis, Univ. of Wisconsin, Madison.

70. ————, and R. F. Harris. 1979. Effect of water potential on the growth and survival of *Escherichia coli* under subsaturating energy substrate conditions. Agron. Abstr. p. 166.

71. ————, A. Unluturk, and R. F. Harris. 1978. Comparative osmotic water potential relations of *Arthrobacter crystallopoietes* and *Escherichia coli*. Agron. Abstr. p. 147.

72. Watson, T. G. 1970. Effect of sodium chloride on steady-state growth and metabolism of *Saccharomyces cerevisiae*. J. Gen. Microbiol. 64:91–99.

73. Wilson, J. M., and D. M. Griffin. 1975a. Water potential and the respiration of microorganisms in the soil. Soil Biol. Biochem. 7:199–204.

74. ————, and ————. 1975b. Respiration and radial growth of soil fungi at two osmotic potentials. Soil Biol. Biochem. 7:269–274.

75. Wolf, A. V., M. G. Brown, and P. G. Prentiss. 1973. Concentrative properties of aqueous solutions: Conversion tables, p. D190–E72. *In* R. C. Weast (ed.) Handbook of chemistry and physics. 54th Edition. CRC Press, Cleveland.

76. Wong, P. T. W., and D. M. Griffin. 1976. Bacterial movement at high water potentials. I. In artificial and natural soils. Soil Biol. Biochem. 8:215–218.

APPENDIX I

Interrelationships Between Water Potential, Water Activity and Concentration for Selected Solutes

The data in Tables A-1 through A-3 were calculated by interpolation using the following equations and constants.

Data based on the concentrative properties of aqueous solutions (20 C) tabulated by Wolf et al. (1973).

$$\psi = -\frac{RT\,D_w\,\overline{O}}{1{,}000} = -24.331\,\overline{O} \quad [1]$$

$$V_s' = V' - V_w' = \frac{10^6}{C_w} - \frac{10^3}{D_w} = \frac{10^6}{C_w} - 1{,}002 \quad [2]$$

$$\mathit{M} = \frac{1{,}000\ \mathrm{M}}{C_w} \quad [3]$$

$$C_s = \mathrm{M}(\mathrm{MW})_s \quad [4.1]$$

$$\overline{C_s} = \mathit{M}\,\mathrm{MW} \quad [4.2]$$

$$A\% = \frac{100\,\mathit{M}(\mathrm{MW})_s}{1{,}000 + \mathit{M}(\mathrm{MW})_s} \quad [4.3]$$

$$C_s\% = \frac{\mathrm{M}(\mathrm{MW})_s}{10} \quad [4.4]$$

where ψ is the water potential (bars: 1 bar = 10^6 dynes•cm^{-2}); R is the universal gas constant (83.1432×10^6 dyne-cm•mole^{-1}•°K^{-1}); T is the temperature (293.16 K for 20 C); D_w is the density of water (0.9982 Kg H_2O•liter H_2O^{-1} at 20 C); $\overline{O}$ is the osmolality (osmoles solute•Kg H_2O^{-1}); V_s' is the specific volume of water displaced by the solute (ml H_2O•Kg H_2O^{-1}), V' is the total specific volume of the solution (ml solution•Kg H_2O^{-1}), and V_w' is the specific volume occupied by the water (1,002 ml H_2O•Kg H_2O^{-1} at 20 C) in a molal solution; C_w is the concentration of water (g H_2O•liter solution^{-1}) in a molar solution; M and *M* are the molarity (moles solute•liter solution^{-1}) and molality (moles solute•Kg H_2O^{-1}), respectively; C_s is the molar-based (g solute•liter solution^{-1}) and $\overline{C_s}$ is the molal-based (g solute•1,000 g H_2O^{-1}) anhydrous solute concentration; A% is the % (W/W)-based (g solute•100 g solution^{-1}) and $C_s\%$ is the % (W/V)-based (g solute•100 ml solution^{-1}) anhydrous solute concentration.

Data based on the osmotic coefficients (25 C) tabulated by Robinson and Stokes (1959).

$$\psi = \frac{RT}{V_w^\circ} \ln A_w = \frac{RT\,D_w \ln A_w}{(MW)_w} = -\frac{RT\,D_w\,v\,M\,\phi}{1{,}000}$$

$$= -24.717\,v\,M\,\phi = -24.717\,\overline{O} \qquad [5]$$

$$M = \frac{1{,}000\,D_w\,M}{1{,}000 + M(MW)_s} = \frac{997.1\,M}{1{,}000 + M(MW)_s} \qquad [6]$$

where A_w is the water activity (solution vapor pressure•pure water vapor pressure^{-1}); V_w° is the molar volume of water (18.0683 ml H_2O•mole H_2O^{-1} at 25 C); D_w is the density of water (0.99707 g H_2O•ml H_2O at 25 C); $(MW)_w$ is the molecular weight of water (18.0154 g H_2O•mole H_2O^{-1}); v is the number of species per molecule of solute; ϕ is the osmotic coefficient (osmolality•ideal osmolality^{-1}).

Data based on the K_2 constants (25 C) derived by Norrish (1966).

$$\psi = \frac{2.303\,RT \log A_w}{V_w^\circ}$$

$$= \frac{2.303\,RT}{V_w^\circ}\left[\log\frac{Mw}{Mw + M} - K_2\left(\frac{M}{Mw + M}\right)^2\right]$$

$$= 3{,}159\left[\log\frac{55.508}{55.508 + M} - K_2\left(\frac{M}{55.508 + M}\right)^2\right] \qquad [7]$$

where Mw is the molality of water (55.508 moles H_2O•1,000 g H_2O^{-1} and K_2 is a proportionality constant for a given solute at a given temperature (25 C).

Basic relationship between ψ and A_w (equation 5).

$$\psi = 1350.573 \ln A_w \text{ (20C)}$$

$$= 1372.012 \ln A_w \text{ (25C)}$$

APPENDIX II

Model for Evaluating the Effect of Water Potential on Cell Composition

The following reference model of the effect of ψ on cell composition builds on the ψ and solute balance schematic (Fig. 1) used to illustrate the different components of microbial substrate and microbial cell ψ as described in the text. Possible ψ_m contributions to the ψ of microbial cell walls and protoplasts and the possibility of solute interactions modifying the additive nature of ψ_s are not considered. It should be emphasized that the major purpose of the model is to establish principles and illustrate the relative magnitude of cell composition changes to specific physiological responses to ψ stress. The composition of the unstressed reference cell is standardized to reflect that of a typical Gr^- procaryote. No attempt has been made to adjust the unstressed cell composition to reflect the wide variation in cell composition shown by different procaryotic and particularly eucaryotic microorganisms, but the principles described are readily applicable to any microorganism. The model evaluates the implication of transient and growth-related turgor pressure changes on cell composition but assumes, for simplicity, that cell wall rigidity is sufficiently high so that cell volume remains essentially constant for stressed vs. nonstressed cells. However, the model is sufficiently flexible to readily allow modification for cell volume fluctuation under ψ stress. Composition in the unstressed state must be established experimentally, and interpretation of the effect of ψ stress on cell composition must include consideration of the effect of associated ψ stress effects on growth, particularly growth rate, on cell composition and morphology.

A typical empirical composition of Gr^- bacterial cells is, recognizing a K^+ content of about 1%, $C_4H_7O_{1.5}NK_{0.03}$ (+9% F.W. other elements = 9.3 g): 103.5 g/mole cells formula weight (F.W.); 46% C; 13.5% N; 1% K; 17 C-bound electron equivalent cells/mole cells (Harris and Adams, 1979). Differentiation of the cell into a polymer matrix (potentially non ψ-dependent from a chemical composition standpoint) and a dissolved solute (potentially highly ψ-dependent) phase identifies, assuming a 6% W/W dissolved solute concentration of which 1% is K^+ and the remaining 5% has the same composition as the intact cell material, a polymer matrix phase contribution of 0.95 ($C_4H_7O_{1.5}N$ + 9.3 g) and a basal solutes phase contribution of 0.05 ($C_4H_7O_{1.5}NK_{0.6}$ + 8.1 g) (Table A-4). As noted previously, for a water content of 2.32 g H_2O/g cells, the basal solute phase can be considered to consist of a 0.2*M* solution with a ψ of −7 bars (Fig. 2; Table A-4). From an intracellular solution volume standpoint, the basal metabolite solutes are dissolved in 2.32 × 103.5 = 240 g = 240.6 ml H_2O/mole cells (Table A-4). The displacement volume of 0.2*M* of a mixture of K^+ salts of amino acids, carboxylic acids and carbohydrates is about 10 ml/1,000 ml H_2O (Table A-1), or 10 × 240.6/1,000 = 2.4 ml/mole cells (Table A-4). Thus, the intracellular solution volume is about 243 ml/mole cells for the unstressed Gr^- bacterial cell (Fig. 1, 2; Table A-4). Due to the displacement volume phenomenon, any ψ-strategy-related accumulation of compatible or stress solutes will either cause an increase in the intracellular solution volume, and thus also the cell volume, or will, more likely, result in displacement of a corresponding amount of water, thereby causing cell dehydration. This dehydration, together with an increase in total cell weight due to accumulation of intracellular ψ-controlling solutes, may result in a substantial decrease in cell H_2O content (W/W) as well as in a substantial change in chemical composition under ψ stress (Fig. 2; Table A-4, A-5). For example, a mechanistic response of induction of −20 bar K^+ glutamate$^-$ and −20 bar proline and accumulation of −40 bar NaCl for growth under −80 bar NaCl stress affects cell composition as follows (Table A-4). The intracellular solution will now be composed not only of the basal metabolites (0.2*M*) but will also contain about 0.45*M* K^+ glutamate$^-$, 0.8*M* proline and 0.9*M* NaCl (Tables A-1 and A-3; Table A-4). Assuming constant cell volume and constant basal metabolite concentration (molal/based), the combined solutes will cause a displacement of about 20 ml H_2O/mole cells, giving a new water content of 219.8 ml = 219 g H_2O/mole cells (Table A-4). The corresponding solute concentration (moles solute/219 g H_2O/mole cells) will be 0.21 × 0.243 = 0.046 moles basal solutes, 0.46 × 0.243 = 0.101 moles K^+ glutamate$^-$, etc. (Table A-4). Relative contributions of the different cell components to cell composition are identified in Table A-4. The decreased water content, 219/153.7 = 1.42 g H_2O/g cells, is due more to the cell dry weight increase from 103 to 154 g/mole than to the 20 ml H_2O displacement per mole cells. The cell empirical composition, standardized to 1 mole cell N for comparison with the non ψ-stressed reference, is $C_{4.67}H_{7.32}O_{1.75}NK_{0.10}(NaCl)_{0.16}$(+7.2 g); 120.1 g/mole F.W; 47% C; 11.7% N, 3.2% K; 17.7 C-bound e^-eq/N. Similar data for different mechanistic ψ responses are summarized in Table A-5 and Fig. 2.

LIST OF SYMBOLS

Symbol	Specification	Most commonly used dimension
A_{ATP}	Total specific (dATP/dX) adenosine triphosphate (ATP) requirement for biosynthesis and/or maintenance, for a given growth rate (growing cells) or for a given time period (non-growing cells)	moles ATP•mass cells^{-1}
A^{b}_{ATP}	Specific ATP requirement for biosynthesis (nominally μ-independent)	moles ATP•mass cells^{-1}
A^{m}_{ATP}	Specific ATP requirement for maintenance, for a given μ (growing cells: $A^{m}_{ATP} = q^{m}_{ATP}/\mu$) or for a given time period (non-growing cells: $A^{m}_{ATP} = q^{m}_{ATP}/\Delta t$)	moles ATP•mass cells^{-1}
A^{m-b}_{ATP}	Specific ATP requirement for maintenance of active biosynthetic capability	moles ATP•mass cells^{-1}
A^{m-v}_{ATP}	Specific ATP requirement for maintenance of cell integrity and viability	moles ATP•mass cells^{-1}
A_{ATPgen}	Specific amount of ATP generated for biosynthesis and/or maintenance; subdivided into A^{b}_{ATPgen}, A^{m}_{ATPgen}, etc.	moles ATP•mass cells^{-1}

Symbol	Specification	Most commonly used dimension
D	Dilution rate for continuous culture	hr^{-1}
Ec	Efficiency of energy coupling between ATP generation and ATP use. expressed as a ratio with respect to E_c^{MAX}	dimensionless
A_s	Total specific (dS/dX) amount of substrate (S, electron donor/organic carbon source, unless specified otherwise) used for biosynthesis and/or maintenance	moles substrate•mass $cells^{-1}$
A_s^b	Specific amount of substrate assimilated and dissimilated for biosynthesis	moles substrate•mass $cells^{-1}$
A_s^m	Specific amount of substrate dissimilated for maintenance; subdivided into A_s^{m-b} and A_s^{m-v}	moles substrate•mass $cells^{-1}$
$A_s^{\mu max}$	Specific amount of substrate used for growth when $\mu = \mu max$	moles substrate•mass $cells^{-1}$
$A_s^{\mu max(sat)}$	Specific amount of substrate used for growth under saturating S conditions (e.g. batch culture)	moles substrate•mass $cells^{-1}$
$A_s^{\mu max(*)}$	Specific amount of substrate used for growth at $S_{min}^{\mu max}$ (e.g. extrapolated continuous culture)	moles substrate•mass $cells^{-1}$
A_{a-s}	Specific amount of substrate assimilated for biosynthesis	moles substrate•mass $cells^{-1}$
A_{a-s}^{es}	Specific amount of substrate assimilated for biosynthesis of essential cell materials	moles substrate•mass $cells^{-1}$
A_{a-s}^{st}	Specific amount of substrate assimilated for biosynthesis of storage and/or ψ stress-related cell materials	moles substrate•mass $cells^{-1}$
A_{d-s}	Specific amount of substrate dissimilated for biosynthesis and/or maintenance; subdivided into A_{d-s}^b and A_s^m	moles substrate•mass $cells^{-1}$
A_{d-s}^b	Specific amount of substrate dissimilated for biosynthesis	moles substrate•mass $cells^{-1}$
A_z	Specific amount of metabolite (Z, substrate/product other than the electron donor/organic C source) consumed or produced as a result of biosynthesis and/or maintenance; subdivided into A_{a-z}, A_{d-z}, A_z^m, etc.	moles metabolite•mass $cells^{-1}$

Symbol	Specification	Most commonly used dimension
F	Proportionality factor relating biomass decay rate (λ) to death rate (γ)	dimensionless
j_s	Specific rate of substrate efflux	moles sub•mass cells^{-1}• hr^{-1}
$K_{s(v)}$	Half saturation affinity constant for substrate uptake (S at which $v_s = 0.5\ v_s^{max}$)	μM substrate
lim	(used as a superscript) denotes S-limiting growth conditions	--
MAX	(used as a superscript) denotes the maximum possible value of a physiological property	--
MIN	(used as a superscript) denotes the minimum possible value of a physiological property	--
N_x	Number of individual cells	number cells
$N_{x(max)}$	Maximum number of individual cells	number cells
N_{xv}	Number of viable cells	number cells
q_{ATP}	Total specific rate of ATP use for biosynthesis and/or maintenance	moles ATP•mass cells^{-1}• hr^{-1}
q_{ATP}^{b}	Specific rate of ATP use for biosynthesis (nominally μ-dependent: $q_{ATP}^{b} = \mu A_{ATP}^{b}$)	moles ATP•mass cells^{-1}• hr^{-1}
q_{ATP}^{m}	Specific rate of ATP use for maintenance (commonly called the maintenance energy coefficient, M_{ATP}) nominally μ-independent: q_{ATP}^{m} = constant)	moles ATP•mass cells^{-1}• hr^{-1}
q_{ATP}^{m-b}	Specific rate of ATP use for maintenance of active biosynthetic capability	moles ATP•mass cells^{-1}• hr^{-1}
q_{ATP}^{m-v}	Specific rate of ATP use for maintenance of cell integrity and viability (commonly called the endogenous rate of ATP metabolism)	moles ATP•mass cells^{-1}• hr^{-1}
q_{ATPgen}	Specific rate of ATP generation; subdivided into q_{ATPgen}^{b}, q_{ATPgen}^{m}, etc.	moles ATP•mass cells^{-1}• hr^{-1}
q_{ATPgen}^{MAX}	Maximum specific rate of ATP generation, a potential primary determinant of μ_{max} under non S-limiting conditions	moles ATP•mass cells^{-1}• hr^{-1}
q_s	Total specific rate of substrate consumption for biosynthesis and/or maintenance	moles sub•mass cells^{-1}• hr^{-1}
q_{a-s}	Specific rate of substrate assimilation for biosynthesis	moles sub•mass cells^{-1}• hr^{-1}

Symbol	Specification	Most commonly used dimension
q^{es}_{a-s}	Specific rate of substrate assimilation for biosynthesis of essential cell materials	moles sub•mass cells^{-1}• hr^{-1}
$q^{es(MAX)}_{a-s}$	Maximum specific rate of substrate assimilation for biosynthesis of essential cell materials, a potential primary determinant of μ_{max} under non S-limiting conditions	moles sub•mass cells^{-1}• hr^{-1}
q^{st}_{a-s}	Specific rate of substrate assimilation for biosynthesis of storage and/or ψ stress-related cell materials	moles sub•mass cells^{-1}• hr^{-1}
q_{d-s}	Specific rate of substrate dissimilation for biosynthesis and/or maintenance; subdivided into q^{b}_{d-s} and q^{m}_{s}	moles sub•mass cells^{-1}• hr^{-1}
q^{m}_{s}	Specific rate of substrate dissimilation for maintenance (commonly called the maintenance energy coefficient, M_s)	moles sub•mass cells^{-1}• hr^{-1}
$q^{\mu max(sat)}_{s}$	Specific rate of substrate consumption shown under saturating S growth conditions	moles sub•mass cells^{-1}• hr^{-1}
$q^{\mu max(MIN)}_{s}$	Minimum specific rate of substrate consumption capable of supporting μ_{max} (analogous to $q^{\mu max(*)}_{s}$ = q_s at $S^{\mu max}_{min}$)	moles sub•mass cells^{-1}• hr^{-1}
q^{m}_{x}	Specific rate of endogenous metabolism expressed as a decay rate constant (λ when $q_s = 0$); subdivided into q^{m-b}_{x} and q^{m-v}_{x}	mass cells•mass cells^{-1}• hr^{-1}
R_{sf}	Rate of substrate supply to the cell	moles sub•hr^{-1}
R_{sq}	Total rate of substrate (electron donor/organic C source, unless specified otherwise) consumption for biosynthesis and/or maintenance	moles sub•hr^{-1}
R^{b}_{sq}	Rate of substrate consumption for biosynthesis	moles sub•hr^{-1}
R^{m}_{sq}	Rate of substrate dissimilation for maintenance	moles sub•hr^{-1}
R_{zq}	Rate of consumption/production of microbial metabolite, Z, resulting from biosynthesis and/or maintenance; subdivided into R^{b}_{zq}, R^{m}_{zq}, etc.	moles sub•hr^{-1}

Symbol	Specification	Most commonly used dimension
R_x	Rate of change of microbial biomass	mass cells • hr^{-1}
R_{xv}	Rate of change of viable microbial biomass	mass cells • hr^{-1}
S	Concentration of substrate (electron donor/organic C source, unless specified otherwise)	*M* substrate
$\overline{S}$	Steady state concentration of substrate in a continuous culture reactor/effluent	*M* substrate
S_o	Initial concentration of S (or influent S concentration for a continuous culture)	*M* substrate
S_t	S concentration after time t	*M* substrate
S_m	S concentration at which $q_s = q_s^m$ and below which growth can no longer be supported	μM substrate
$S_{min}^{\mu max}$	Minimum S concentration supporting μ_{max}: below $S_{min}^{\mu max}$, $\mu < \mu_{max}$ is limited by v_s; above $S_{min}^{\mu max}$, $\mu \not> \mu_{max}$ is limited by $q_{a-s}^{es(MAX)}$ and/or q_{ATPgen}^{MAX}	μM substrate
surv	(used as a superscript) denotes a survival mode of metabolism	--
ΔS	Total amount of substrate consumed for production of ΔX biomass	moles substrate
ΔS_a	Amount of substrate assimilated for production of ΔX biomass	moles substrate
ΔS_d	Amount of substrate dissimilated for production of ΔX biomass	moles substrate
t_d	Biomass doubling time	hour
$t_{0.5}$	Biomass half life	hour
T_D	Culture detention or replacement time for continuous culture	hour
v_s	Specific rate of substrate uptake	moles sub • mass $cells^{-1}$ • hr^{-1}
v_s^{max}	Maximum specific rate of substrate uptake with respect to S	moles sub • mass $cells^{-1}$ • hr^{-1}
v_s^{MAX}	Maximum possible specific rate of substrate uptake, a potential primary determinant of μ_{max} under saturating S conditions	moles sub • mass $cells^{-1}$ • hr^{-1}
$v_s^{\mu max(MIN)}$	Minimum specific rate of substrate uptake capable of supporting μ_{max} (analogous to $v_s^{\mu max(*)} = v_s$ at $S_{min}^{\mu max}$)	moles sub • mass $cells^{-1}$ • hr^{-1}

Symbol	Specification	Most commonly used dimension
X	Concentration of microbial biomass	mass cells
$X_{max(s)}$	Maximum biomass concentration for a given amount of S	mass cells
X_o	Initial biomass concentration	mass cells
X_t	Biomass concentration after time t	mass cells
X_v	Concentration of viable biomass	mass cells
$X_{v(o)}$	X_v at time zero	mass cells
$X_{v(t)}$	X_v at time t	mass cells
X^{gr}_{max}	Maximum biomass of cells that can be supported by a given R_{sf} in a growth mode of metabolism	mass cells
X^{surv}_{max}	Maximum biomass of cells that can be supported by a given R_{sf} in a survival mode of metabolism	mass cells
Y_s	Specific growth yield of cells (dX/dS) for a given μ	mass cells•mole substrate^{-1}
Y^b_s	Specific yield of cells assuming all the S was used for biosynthesis (Y_s at $\mu = \infty$ so that $A^m_s = 0$) (commonly called Y^{max}_s)	mass cells•mole substrate^{-1}
$Y^{\mu max}_s$	Specific growth yield for cells growing at μ_{max}	mass cells•mass substrate^{-1}
$Y^{\mu max(sat)}_s$	Specific growth yield under saturating S growth conditions (e.g. batch culture)	mass cells•mole substrate^{-1}
$Y^{\mu max(*)}_s$	Specific growth yield at $S^{\mu max}_{min}$ (e.g. extrapolated continuous culture)	mass cells•mole substrate^{-1}
Z	Concentration of metabolite (other than the electron donor/organic C source)	*M* metabolite
Z_t	Concentration of Z after time t	*M* metabolite
$Z_{max(s)}$	Maximum concentration of Z resulting from a given S addition	*M* metabolite
$\alpha_{z/s}$	Proportionality factor relating Z to S	moles metabolite•mole substrate^{-1}
γ	Specific death rate of biomass $(-dX_v/dt)$	mass cells•mass cells^{-1}•hr^{-1}
$\xi_{ATP/s}$	Specific ATP yield per unit mass of substrate dissimilated	moles ATP•mole substrate^{-1}
$\xi_{ATP/x}$	Specific ATP yield per unit mass of cell material dissimilated	moles ATP•mass cells^{-1}
$\xi_{ATP/z}$	Specific ATP yield per unit mass of Z dissimilated	moles ATP•mole metabolite^{-1}
$\xi_{s/x}$	Specific energy equivalenct of dissimilated substrate versus dissimilated cell material	moles substrate•mass cells^{-1}

Symbol	Specification	Most commonly used dimension
λ	Specific decay rate of biomass $(-dX/dt)$	mass cells•mass cells^{-1}•hr^{-1}
μ	Specific growth rate of biomass (dX/dt)	mass cells•mass cells^{-1}•hr^{-1}
μ_{max}	Maximum specific growth rate	mass cells•mass cells^{-1}•hr^{-1}
μmax	(used as a superscript) denotes $\mu = \mu_{max}$ for a μ-dependent physiological property	--
Θ	Cell viability index (probability that a cell will be viable at birth)	dimensionless
Θ_w	Gravimetric soil water content (W/W)	g H_2O•g dry soil^{-1}
ω	Gravimetric cell water content (% W/W)	g H_2O•g dry cells^{-1}
χ	Specific biomass of a cell	mass•cell^{-1}
*	(used as a superscript) denotes the value of a physiological property shown at $S_{min}^{\mu max}$	--
ψ	Water potential	bars
ψ_{cells}	Total ψ of cells	bars
$(\psi_m)_{cells}$	Cell ψ due to water sorption by intracellular colloid and particulate surfaces (bound water)	bars
$(\psi_p)_{cells}$	Cell turgor pressure component arising from establishment of hydrostatic pressure across the cell membrane due to $(\psi_m + \psi_s)_{cells} < \psi_{sub}$	bars
$(\psi_s)_{cells}$	Cell ψ due to water sorption by intracellular solutes	bars
$(\psi_{bs})_{cells}$	Cell ψ_s due to basal intermediary metabolites	bars
$(\psi_{cs})_{cells}$	Cell ψ_s due to constitutive compatible solutes	bars
$(\psi_{is})_{cells}$	Cell ψ_s due to inducible compatible solutes	bars
$(\psi_{ss})_{cells}$	Cell ψ_s due to stress solutes present in high concentrations in the substrate	bars
ψ_{sub}	Total ψ of substrate	bars
$(\psi_m)_{sub}$	Substrate ψ due to water sorption by particulate surfaces, colloids and solutes impermeant with respect to bacterial cell wall/membranes	bars

Symbol	Specification	Most commonly used dimension
$(\psi_s)_{sub}$	Substrate ψ due to water sorption by solutes permeant (by active and/or passive mechanisms) with respect to bacterial cell wall/membranes (commonly called the osmotic water potential)	bars
$(\psi_{bs})_{sub}$	Substrate ψ_s due to basal solutes	bars
$(\psi_{ss})_{sub}$	Substrate ψ_s due to stress solutes	bars

CHAPTER 3

The Effect of Water Potential on Decomposition Processes in Soils[1]

L. E. SOMMERS, C. M. GILMOUR, R. E. WILDUNG, AND S. M. BECK[2]

INTRODUCTION

The importance of soil-water relationships as a significant factor in the decomposition of organic matter in soils has been recognized for many years. Qualitatively, this has been shown by the higher organic matter content of poorly-drained soils when compared with their well-drained counterparts, and the accumulation of organic matter in permanently saturated environments such as lake sediments and marshes. As discussed by Griffin (1972), numerous experiments have been conducted to evaluate the influence of water content on microbial processes in soils. However, it is difficult to separate the direct effects of water potential from secondary effects (e.g., aeration, solute diffusion) and other parameters. Furthermore, the results obtained in many studies have been expressed in non-comparable terms, i.e., water content, percentage of permanent wilting point, or percentage of water holding capacity, which cannot be related for various soils unless the relationship between water content and water potential is known. The objective of this paper will be to summarize the limited information available on the effect of water potential on the degradation in soil of organic compounds including soil organic matter, plant residues, synthetic organic chemicals, and waste materials, and to present approaches developed for modeling the influence of soil moisture on decomposition processes.

[1]Joint contribution from the Indiana Agric. Exp. Stn. (J. Paper No. 7894), Idaho Agric. Exp. Stn. (J. Paper No. 79516) and Batelle Pacific Northwest Laboratories.

[2]Professor of agronomy, Purdue Univ., W. Lafayette, IN 47907; professor of bacteriology & biochemistry, Univ. of Idaho, Moscow, ID 83843; manager, environmental chemistry, Batelle Pacific Northwest Laboratories, Richland, WA 93352 and; professor of bacteriology and biochemistry, Univ. of Idaho, Moscow, ID 83843.

DECOMPOSITION PROCESSES

The basic decomposition process in soils involves a sequential, microbial conversion of reduced organic C compounds to an oxidized end-product, principally CO_2, under aerobic conditions, and to a variety of incompletely oxidized C compounds when O_2 levels are depressed. The rate of decomposition of organic materials in soil is influenced by temperature, O_2 supply, water potential, pH, inorganic nutrients, and the C:N ratio of the material along with management factors such as tillage.

To evaluate the influence of soil water potential on the product of decomposition, it is essential to separately examine soil systems possessing unsaturated aerobic conditions (i.e., potentials < -0.2 to -0.1 bars). The organisms involved in decomposition under aerobic conditions include bacteria, fungi, actinomycetes, and various soil microfauna. At soil water potentials within the range encountered for growth of higher plants (-0.3 to -15 bars), microbial metabolism involves the aerobic oxidation of organic C to CO_2. With decreasing water potentials, the activity of bacteria appear to be limited first, mainly because the decreasing proportion of water filled pores results in reduced mobility of bacterial cells and limited availability of substrates (i.e., reduced diffusion of solutes). As discussed by Griffin (1978), actinomycetes and many of the fungi are capable of surviving at very low water potentials (e.g., of -40 to -100 bars), but may be metabolically inactive.

The relationship between soil microorganisms and the soil water potential is governed not only by each organisms ability to function as the available water fluctuates from one extreme to another, but also by the type of substrate, pH, temperature, nutrients, and the ability to compete with other organisms.

As the water content of soils increases, the percentage of air filled pores decreases resulting in a decreased O_2 tension in the soil and the development of anaerobic conditions. Concurrently, the microbial population changes from predominantly aerobic organisms to facultative anaerobic organisms and, with less O_2, obligate anaerobic organisms. The increased water content also enables greater solute diffusion to cells and movement of microbes from one microenvironment to another. The microbial population developing under anaerobic conditions consists of numerous genera of bacteria, resulting in the formation of diverse metabolic end-products during decomposition of organic compounds. For example, immediately after onset of anaerobiosis, facultative anaerobes oxidize organic C to primarily CO_2, whereas equivalent amounts of CH_4 and CO_2 are produced at subsequently lower O_2 tensions, under appropriate substrate and pH conditions. At intermediate O_2 concentrations, fermentation reactions may result in the production of organic acids, alcohols and other partially oxidized organic C compounds. The terminal electron acceptor utilized by bacteria shifts as the soil system becomes progressively more reduced. The following sequence of electron acceptors are typically utilized following the onset of anaerobiosis in soils containing a supply of oxidizable organic C; 1) O_2, 2) NO_3^-, 3) Mn^{4+} (MnO_2), 4) Fe^{3+} (Fe^{3+} in Fe_2O_3 and other iron oxides), 5) SO_4^{2-}, 6) H_2, and 7) CO_2. The utilization of these compounds as terminal electron acceptors results in the formation of gaseous products characteristic of anaerobic systems such as CH_4, N_2, N_2O, H_2, and H_2S. Various aspects of microbial reactions occurring in flooded soils have been summarized by Yoshida (1975). The majority of constituents contained within plant, animal, and microbial residues are partially degraded to incompletely oxidized organic C compounds under anaerobic conditions. However, some organic constituents, such as aromatic compounds, are not appreciably altered under anaerobic conditions because their degradation involves direct incorporation of O_2 into the aromatic ring to achieve structural cleavage. Metabolism of the degradation products through the tricarboxylic acid (Krebs) cycle also requires O_2.

DECOMPOSITION OF SOIL ORGANIC MATTER AND RESIDUES

The decomposition of soil organic matter will be assessed in terms of two different processes: 1) CO_2 evolution and 2) nitrogen mineralization. Water potential and temperature are major factors influencing both processes.

Several studies have been conducted to evaluate the influence of osmotic water potential on the decomposition of soil organic matter. The addition of Na_2SO_4, NaCl, or a salt mixture to soils appears to reduce CO_2 evolution in a non-linear fashion over the -0.5 to -30 bar range (Singh et al., 1969; Johnson and Guenzi, 1963). Decreasing the matric water potential results in a linear decrease in the amount of CO_2 evolved from decomposition of soil organic matter (Miller and Johnson, 1963). Optimum water potentials for decomposition of soil organic matter are within the -0.2 to -0.5 bar range. Evolution of CO_2 as a function of water potential is shown for a semi-arid grassland site in Fig. 1 (Wildung et al., 1975). The evolution of CO_2 decreased rather rapidly between -0.3 and -15 bars and then decreased linearly to approximately -80 bars. There was a strong dependence of CO_2 evolution on both soil temperature and water potential. The relationship between CO_2 evolution and water potential was a function of temperature. A simple regression equation, which included a soil temperature-water interaction term, accounted for a major portion of the variation in respiration rate over a 2 year period. When temperatures were suboptimal (6 to 12 C) for decomposition, water potential had less influence on CO_2 evolution than when more optimum temperatures (18 to 24 C) prevailed. This dependence on temperature suggests that changes in CO_2 evolution resulted from distinct changes in microbial types, numbers, and metabolic activities.

Proper management of crop residues requires a detailed knowledge of the rate of decomposition of various plant tissues over the range of conditions encountered in the field. The interaction between soil temperature and water potential in the decomposition of Blue grama residues has been evaluated in detail by Nyhan (1976). At a soil water potential of -115 bars, a linear increase was found for percent carbon loss and soil temperature over the 2 to 60 C range. However, at more optimal soil water potentials (e.g., -0.01 to -5.9 bars), the maximum rate of CO_2 evolution was found to occur between 20 and 35 C. At a constant temperature, de-

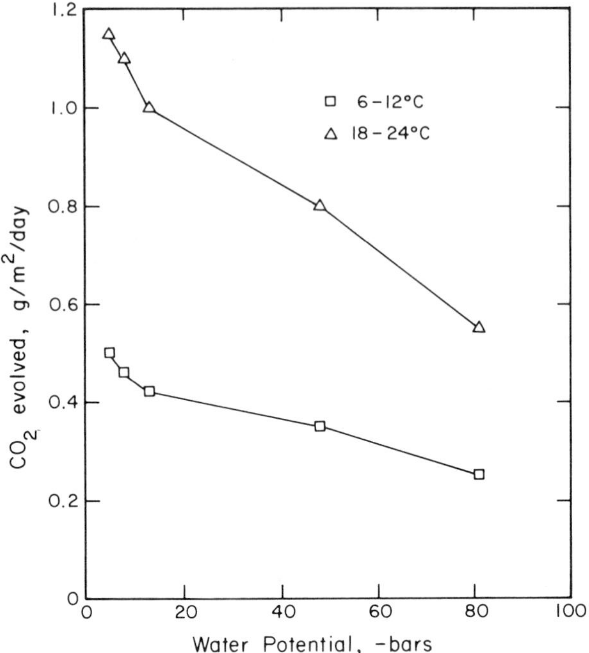

Fig. 1. Evolution of CO_2 from an undisturbed shrub steppe soil at different temperatures during the year (Wildung et al., 1975).

creasing soil water potential from −0.01 bars to −100 bars resulted in a 100 to 1,000 fold decrease in the rate of CO_2 evolution. These data indicate the importance of either controlling (lab studies) or measuring (field studies) soil temperatures when attempting to evaluate the influence of soil water potential on decomposition in soils.

Under natural conditions, and in some agricultural situations, plant residues may not be directly incorporated into the soil. Early studies were conducted to evaluate the decomposition of various plant residues in the absence of soils (Bartholomew and Norman, 1946). The results of such investigations (Fig. 2) indicated that significant amounts of most residues are decomposed at water potentials considerably lower than those commonly occurring in soils. For example, the water potential at which CO_2 evolution was decreased by 50% of maximum was at approximately −40 bars. Decomposition was detectable at water potentials of approximately −200 bars. It is apparent that decomposition of organic residues in soils will occur over the entire range of soil water potentials encountered. With high residue crops, it is often desirable to optimize the decomposition rate. In a study of five soils maintained at relatively high water potentials, the maximum initial rate of corn stover decomposition occurred at approximately −0.05 bars (Bhaumaik and Clark, 1947). These results were obtained over a relatively short incubation period (15 days) and are rather typical of other experiments in that optimal decomposition of plant residues occurs at approximately −0.3 bars or 60% of the water holding

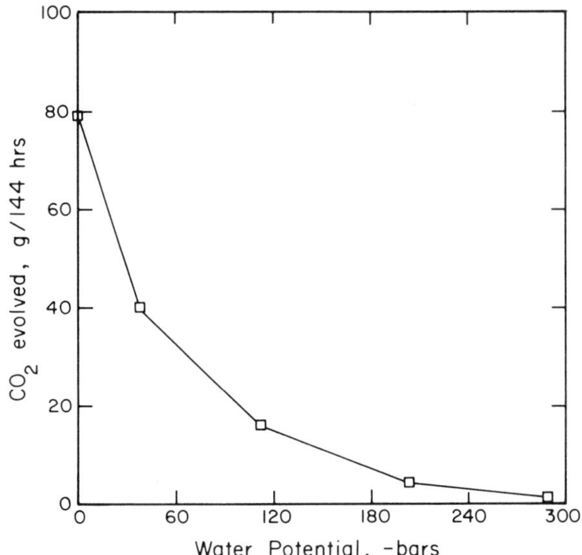

Fig. 2. Effect of water potential on decomposition of oat straw in the absence of soil (Bartholomew and Norman, 1946).

capacity (WHC). The alteration of the osmotic potential of soils through addition of NaCl and $CaCl_2$ in conjunction with addition of plant residues indicates that a relatively rapid decrease in CO_2 evolution occurs between -0.05 to -10 bars followed by a linear decrease to approximately -100 bars (Laura, 1974). The effects of various management practices on the decomposition of rice straw has indicated that maximum decomposition occurred at 30% of the WHC (~ -0.3 bars) and was decreased at either 60 or 150% of the WHC (Pal and Broadbent, 1975; Sain and Broadbent, 1976). In a study designed to compare the decomposition of rice straw under cyclic aerobic and anaerobic conditions, as opposed to continuous aerobic or anaerobic incubation, Reddy and Patrick (1975) found that the extent of decomposition decreased as the length of anaerobic period increased (Table 1). After 128 days of incubation, CO_2 production was reduced by more than 50% when soils were maintained under continuous anaerobic conditions as compared with aerobic conditions. Methane was not found to be a significant end-product of decomposition in these systems.

To determine the fraction of plant residues decomposed, it is necessary to distinguish between CO_2 evolution from decomposition of native soil organic matter and that originating from degradation of residues. This requires either analysis of soils for residual organic C (Wildung et al., 1975) or use of ^{14}C-labeled tissues. Table 2 has a summary of recent research utilizing ^{14}C-labeled plant materials in field experiments to obtain estimates on the extent of decomposition. Concurrent with measurements of $^{14}CO_2$ loss from plant materials incubated in the field, other variables such as temperature and moisture are often measured to obtain

Table 1. Decomposition of rice straw as influenced by alternate aerobic and anaerobic conditions.†

Length of incubation per cycle			
Aerobic	Anaerobic	Number of cycles	CO_2-C evolved‡
——— days ———			mg/g
2	2	32	3.0
4	4	16	3.2
8	8	8	2.8
16	16	4	2.9
32	32	2	2.8
64	64	1	2.5
--	128	--	1.4
128	--	--	3.2

† Adapted from Reddy and Patrick, 1975.
‡ Crowley silt loam amended with 0.5% rice straw (48% C).

Table 2. Decomposition of ^{14}C labeled plant materials in field experiments.

Material	Time	^{14}C lost	Reference
	years	%	
Blue grama herbage	1.1	54–57	Nyhan, 1975
roots	1.1	26–37	
Ryegrass	1.0	67	Jenkinson, 1965
Corn stover	1.0	53–67	Oberlander et al., 1968
Wheat straw	1.0	69	Fuhr and Sauerbeck, 1968
Wheat straw	1.0	55–74	Shields and Paul, 1973
	3.0	79–85	

better correlations with the rates of decomposition. Typical results from field experiments using ^{14}C labeled plant materials indicates that approximately 50 to 75% of the organic carbon in various residues will be decomposed during the first year after addition to soil.

To summarize the influence of water potential on decomposition of soil organic matter and crop residues, several sets of data were obtained from the literature and expressed on a common base to facilitate comparisons. The incubation times, soil properties, and methods of adjusting soil-water potential vary for the studies reported. Thus, all data were expressed on a relative basis such that the rate (or amount) of CO_2 evolution occurring at a given water potential was divided by the optimum rate (or amount) found in that particular study. Since this ratio is consistent within each experiment, it represents relative rates of decomposition according to the following expression:

$$k/k_{max} \propto \psi$$

where k = decomposition rate at a specific water potential; k_{max} = decomposition rate at optimum water potential; and ψ = water potential, − bars. The sources of experimental data are summarized in Table 3. Since decomposition typically decreases as water content approaches saturation (Bhaumik and Clark, 1947; Reddy and Patrick, 1975), only

Table 3. Experimental data employed in evaluating organic matter decomposition-water potential relationships.

Reference	Method of controlling ψ	Incubation time	Soil amendment
Johnson and Guenzi, 1963	Osmotic	20	None
Miller and Johnson, 1964	Matric	14	None
Bhaumik and Clark, 1947	Matric	15	Corn stover
Singh et al., 1969	Osmotic	28	None
Laura, 1977	Osmotic†	30	Gulmohur leaves
		180	

† Electrical conductivity measurements in saturation extract were converted to water potential by *assuming* that conductivity of saturation extract was 50% of value in soil at 60% water holding capacity.

CO_2 evolution data obtained at $\psi < -0.05$ were used in developing the regression equation shown in Fig. 3. A relationship significant at the 95% level was obtained between k/k_{max} and the log of water potential ($r = 0.77$, 109 data points) resulting in the following equation:

$$k/k_{max} = 1.164 \log \psi + 0.871$$

The water potential must be converted to a positive value for use in the above equation. The above relationship is applicable to both decomposition of soil organic matter and crop residues added to soils.

NITROGEN MINERALIZATION

The relatively constant C:N ratio of soil organic matter (approximately 10:1) provides a reliable index for evaluation of soil organic matter decomposition based on the rate and extent of N mineralization. Experimental data reported by Stanford and Epstein (1974) were converted from water content to water potential and the relative rates of N mineralization (k/k_{max}) were calculated. Figure 4 shows relative rates of N mineralization are related logarithmically to soil-water potential in a manner similar to that shown above for CO_2 evolution. A linear relationship was found by Stanford and Epstein (1974) between N mineralized and soil water content. Decreases in mineralization are observed in some soils, particularly clays, at high water contents ($\psi > -0.1$ bars) and these data were not included in Fig. 4. A possible explanation for reduced N mineralization at high water contents is that N mineralization and nitrification would proceed rapidly in the aerobic microenvironments and might then be followed by diffusion of NO_3^- into anaerobic microsites where denitrification would occur, resulting in a net loss of N from the system. Additional N mineralization studies have been conducted where salt solutions have been added to soils to alter the osmotic component of the water potential. In essence, these studies indicate that N mineralization can be reduced by lowering the osmotic component of soil water potential (Laura, 1977; Robinson, 1957; Wetselaar, 1968; Reichman et al., 1966; Miller and Johnson, 1964). In soil systems treated with salts, there does not appear to be any differential effect on N mineralization by NaCl, $CaCl_2$,

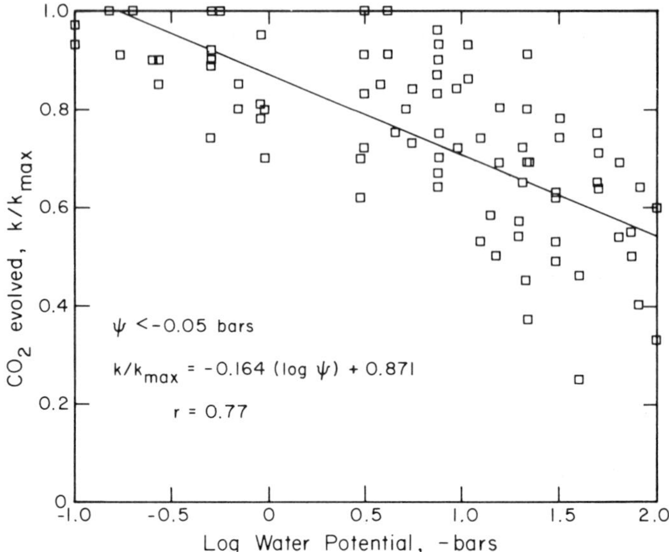

Fig. 3. Relationship between CO_2 evolved, expressed as a relative rate, and soil water potential. See Table 3 for sources of data.

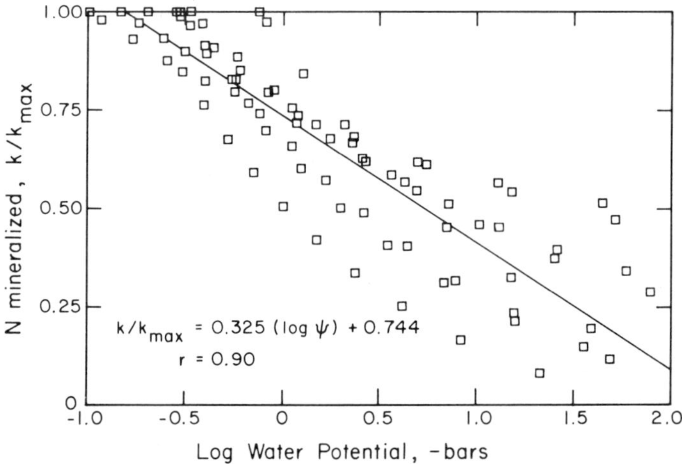

Fig. 4. Relationship between relative rates of N mineralization (k/k_{max}) and soil water potential. Based on data of Stanford and Epstein (1974).

and Na_2SO_4, $CaSO_4$, KCl, $MgCl_2$, K_2SO_4, and $MgSO_4$ when added to soils at varying rates (Sindhu and Cornfield, 1967b). The effect of both matric and osmotic potential on N mineralization has been evaluated under laboratory conditions (Sindhu and Cornfield, 1967a). Minimal amounts of N mineralization were detected in systems maintained under saturated conditions while the optimum water potential for N mineralization occurred between −0.3 to −3 bars and decreased at −5 bars (Fig. 5). The

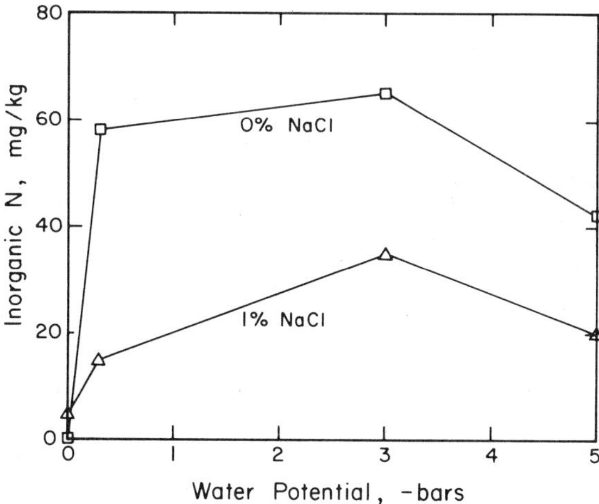

Fig. 5. Effect of matric and osmotic water potentials on N mineralization in soils (adopted from Sindhu and Cornfield, 1967a).

addition of NaCl to soils resulted in a depression of N mineralization at all matric water potentials. Unfortunately, the soil water content was not stated and thus, it is not possible to convert the salt concentration shown in Fig. 5 to an osmotic water potential.

DECOMPOSITION OF WASTE MATERIALS

Currently there is considerable interest in the utilization of waste materials on soils, resulting in the need for information on the relationship between the rates of decomposition and soil water potential. Relatively few studies have been done to evaluate the decomposition of wastes over a wide range of soil moisture conditions. The decomposition rate of sewage sludge organic C was found to be essentially the same over the −0.25 to −1 bar range (Terry et al., 1977). In another study (Sommers et al., 1979), CO_2 evolved from sludge-amended soils was compared at −0.3 bars and 0 bars (saturation). After a 1-year incubation period, CO_2 evolution was 50% lower under saturated conditions than at −0.3 bar water potential. Similar results were obtained by Miller (1974) when three soils ranging in texture from a sand to a clay were incubated for 180 days under saturated conditions and at −0.3 bar water potential. Field respirometry studies have indicated that the rate of feedlot manure decomposition decreases by 50% when soil water decreases from 60 to 30% of WHC (Gilmour et al., 1977).

More recently, C. M. Gilmour and S. M. Beck (unpublished data) used a Palouse silt loam (PSL) and a Ritzville loam (RL) amended with sewage sludge to assess the effects of soil moisture on decomposition of sewage sludge. Soil water contents were established at 0.06, 0.09, 0.12, 0.18, 0.24, 0.30, and 0.36 H_2O/g soil and each soil amended with 1% sewage sludge of known C content. Carbon dioxide evolved was measured over a 30-day period at 26 C. Simple regression analysis was carried out to establish the nature of cause and effect relationships between respiration and water content and/or soil water potential. The value k/k_{max} was used to describe the relative rate of decomposition or microbial activity with k_{max} representing optimal activity. Regression analysis employed both water content (θ) and water potential (ψ). The correlation coefficients (r^2) obtained from the regression of relative decomposition rate and the logarithm of water potential (Fig. 6) were 0.926 and 0.987 for the PSL and RL, respectively. When water content and relative rates of decomposition were regressed, the PSL and RL soils gave correlation coefficients of 0.926 and 0.987, respectively. With these linear regression data on hand, it then becomes feasible to obtain a corrected rate constant based on decreases and/or increases in soil water content and on soil water potential. The equation for half life ($t_{1/2}$) of organic C was modified for each soil as follows:

$$\text{Palouse silt loam } t_{1/2} = \frac{0.693}{k_{max}^{(3.169\,\theta + 0.0054)}}$$

$$\text{Ritzville loam } t_{1/2} = \frac{0.693}{k_{max}^{(3.001\,\theta + 0.1396)}}$$

In a similar way, the water potential regression equation can be used as k_{max} coefficients where:

$$\text{Palouse silt loam } t_{1/2} = \frac{0.639}{k_{max}(0.263 \log \frac{1}{\psi} + 0.611)}$$

$$\text{Ritzville loam } t_{1/2} = \frac{0.693}{k_{max}(0.3119 \log \frac{1}{\psi} + 0.5768)}$$

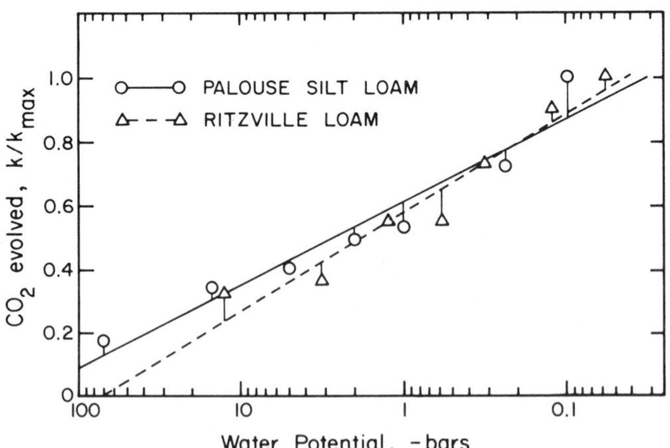

Fig. 6. Relationship between relative rates of sewage sludge decomposition (k/k_{max}) and soil water potential (Gilmour and Beck, unpublished data).

The effect of water potential on half-life values for sludge organic C are summarized in Table 4. The half-life increases with decreasing water content or water potential. However, the effect of soil water is not equivalent in both soils indicating that factors other than soil water potential are of significance in controlling rates of decomposition.

DEGRADATION OF PESTICIDES

As in the case of crop residues, the degradation of pesticides in surface soils is a function of pesticide structure and initial concentration, soil texture, pH, temperature, water potential, and levels of organic matter and other nutrients (Hamaker, 1972; Edwards, 1972; Alexander, 1968). These parameters have complex interdependent effects on the rate of decomposition and the type of metabolic or degradation products formed.

In general, soil properties and environmental conditions which result in increased microbial growth and activity increase the rate of pesticide degradation. Thus, providing the pesticide is not present in concentrations toxic to the microflora, the relationships between water potential and residue decomposition described in previous sections are generally valid for persistence of the intact pesticide molecule. However, in contrast to most natural residues, which are usually innocuous and metabolized through established biochemical pathways, the pesticide molecule and possibly toxic intermediate degradation products may accumulate. This results from the possible absence of enzyme systems necessary for further degradation of a number of unique compounds at any point in the pathway (Alexander, 1967). It is in this area that water potential may have a profound influence through its effects on the types and numbers of soil microorganisms and their metabolic activities.

Water potential will have a direct effect on the types and numbers of soil microorganisms responsible for pesticide degradation. However, assuming adequate soil water for existence of a mixed population, the principal effect of water potential on pesticide degradation pathways arises from the secondary effect on O_2 concentration. It has been sug-

Table 4. Half-life values of sludge organic C at variable soil water contents and corresponding water potentials.

	Half-life based on			
	Water content, g H_2O/g		Water potential, − bars	
Water content	Palouse sil	Ritzville l	Palouse sil	Ritzville l
g H_2O/g soil	half-life, days			
0.06	1,146	1,180	1,783 (−70)†	1,581 (−12)†
0.09	771	920	743 (−15)	900 (− 3)
0.12	581	754	525 (− 5)	688 (− 1)
0.15	466	639	422 (− 2)	582 (− 0.6)
0.18	389	554	367 (− 1)	517 (− 0.3)
0.24	293	438	290 (− 0.02)	439 (− 0.1)
0.30	235	362	251 (− 0.01)	392 (− 0.06)

† Soil water potential, bars, shown in parenthesis (Gilmour and Beck, unpublished).

gested (Alexander, 1968) that increased pesticide persistence under anaerobic conditions may result from the need for molecular O_2 in critical enzymatic reactions. For example, O_2 is required in microbial cleavage of aromatic rings, and there is no known instance of microbial degradation of an aromatic ring under strictly anaerobic conditions in soil.

Much is known of the influence of O_2 concentrations on the gaseous end-products of decomposition in soil and on pathways of decomposition of simple organic compounds, in vitro, e.g., glucose (Stotzky, 1972), and plant residues. However, there are over 1,000 herbicides, insecticides, and fungicides currently registered in the United States, and there is a paucity of information on their degradation pathways in soils. Pesticide metabolism in soil and of known segments of the degradation pathways for major classes of compounds have been reviewed (Alexander, 1967; Kearney and Kaufman, 1972; Meikle, 1972; Edwards, 1972). There is little information concerning the effects of changes in water potential and O_2 concentration on soil pesticide degradative pathways. However, one of the most persistent, and most studied pesticides in soil, DDT[1,1,1-trichloro-2,2-bis(p-chlorophenyl)ethane], serves as an excellent example of the influence of water potential on metabolic pathways.

DDT is relatively resistant to both microbial and chemical degradation in soil. Although a number of metabolites have been determined in biological systems (Wedemeyer, 1967; Brooks, 1974), principal metabolites determined in soil systems (Fig. 7) are DDE[1,1-dichloro-2,2-bis(p-chlorophenyl)ethylene] and DDD[1,1-dichloro-2,2-bis(p-chlorophenyl) ethane]. Under sterile, aerobic conditions, 75% of the initial DDT added

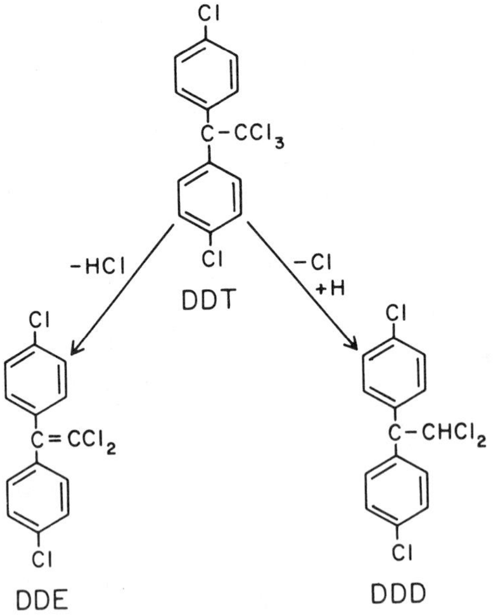

Fig. 7. Initial steps in the dechlorination of DDT in soil. DDD is formed principally through chemical mechanisms whereas DDD is formed as a result of microbial activity. The direction and rate of degradation are markedly influenced by soil aeration and water content.

to a silt loam soil was recovered after incubation for periods up to 24 weeks (Guenzi and Beard, 1968). In nonsterile soils incubated under aerobic conditions, 71.9% of the intact DDT remained and only 2.3% DDE and a trace of DDD were detected after 12 weeks incubation (Table 5). The phenyl rings remained essentially uncleaved and decomposition was minimal in either sterile or nonsterile soils. A similar situation existed for sterile soils incubated under anaerobic conditions (atmosphere of 20% CO_2, 80% N_2). However, in anaerobic, nonsterile systems, less than 15% of the added DDT was recovered after incubation for 12 weeks, and approximately 50% of the DDT added was converted to DDD (Table 5). Amending the nonsterile soil with a readily available C source (1% alfalfa), markedly increased DDT conversion to DDD, with less than 1% intact DDT remaining after incubation. As might be expected from earlier in vitro studies of phenyl ring degradation under anaerobic conditions, there was no evidence of ring cleavage. Concurrently, Plimmer et al., (1968) concluded that DDT undergoes direct reductive dechlorination to DDD without intermediate formation of DDE. Later studies (Parr and Smith, 1974) showed that DDT conversion to DDD in a peaty muck was also accelerated under anaerobic flooded conditions, but total conversion after 4 weeks incubation was considerably lower. Further studies by Guenzi and Beard (1975) verified the relatively slow rate of DDT conversion to DDE and demonstrated that it was principally a chemical process which occurred at a much more rapid rate in submerged soil and in soil at -0.3 bar water potential as compared with air-dry soil. The influence of soil water in this case may be related to sorption/desorption phenomena and increased DDT diffusion to catalytic sites on the soil particle or increased DDT in solution for chemical reaction.

MODELING SOIL MOISTURE AND RATES OF DECOMPOSITION

Although the overall influence of moisture on microbial activity is well recognized, few investigators have attempted to describe decomposition rates as a function of soil water content, water potential or water activity. This is not surprising in view of the complexity of the problem.

Table 5. DDT and degradation products extracted from nonsterile soil incubated under aerobic and anaerobic conditions.†‡

	Recovery after 12 weeks incubation§			
	Without amendment		With 1% alfalfa amendment	
Product	Aerobic	Anaerobic	Aerobic	Anaerobic
	μg			
DDT	71.9	13.6	73.1	0.6
DDE	2.3	0.14	1.7	0.3
DDD	0.3	33.0	0.5	41.0

† Abstracted from Guenzi and Beard (1968).
‡ Over 75% of DDT was recovered from sterile soils, no other products were detected.
§ DDT initially added at a level of 100 μg.

For example, different water relationships (fungi vs. eubacteria), microbial successions, and water transport mechanisms (passive diffusion vs. energy-linked active transport) unique to the protists are difficult to correlate.

The primary questions relate to quantification of the water potential effect. Is it possible that under a set of defined experimental conditions, that organic matter or soil water could be assigned some numerical value that would express quantitative increases or decreases in the rates of substrate décomposition? The following section will attempt to address this question.

At a relatively early date, several investigators (Jenny et al., 1949; Greenland and Nye, 1959; and Olson, 1963) defined the constant k as the rate constant for decay and build-up of organic litter and soil humus. For example, k values are usually expressed in terms of the organic fraction decomposing per unit of time. Thus, a k value of 0.001 day^{-1} indicates that the material is decomposing at the rate of 0.1%/day. Apart from moisture, other variables will modify a particular k value. As a general rule, k values decrease as decomposition proceeds, primarily because of the initial rapid utilization of low molecular weight (water soluble) components and the more gradual hydrolysis and oxidation of complex C-N fractions. In actuality, k is a dynamic parameter rather than a true constant and this fact must be considered when evaluating moisture effects. Hunt (1977) assigned a separate rate constant to the labile (k) and resistant (h) fractions of the organic substrate. Gilmour et al., (1977) used a k value which characterized the decomposition rate approximating a steady state. Paul and VanVeen (1978) noted that it is important to differentiate actual vs. apparent decomposition rates when data are available on recycling of carbon via biomass growth. Other examples could be drawn from the literature to demonstrate that although the k value stands as an excellent theoretical index of reaction velocities, its application to environmental studies remains difficult and should be done only on the basis of defined soil and biological parameters.

The k value becomes an integral part of the conventional exponential decay equation. In this regard, Hunt (1977) discussed departures from the exponential model (e.g., heterogeneous substrates) primarily on the basis of changing rate constants typical of entire decomposition sequences. Most investigators now consider organic matter decomposition in terms of first order rate kinetics (e.g., Paul and VanVeen, 1978; Gilmour et al., 1977) which result in the following expressions:

$$V_{dec} = \frac{dA}{dt} = -kA \qquad [1]$$

where V_{dec} = decomposition rate
k = decomposition rate constant ($time^{-1}$)
A = concentration of added organic matter
t = time

Integration of Eq. 1 yields:

$$A = A_o \cdot e^{-kt} \quad [2]$$

or

$$\ln A/Ao = -kt \quad [3]$$

where A_o = concentration of A at t = 0
Gilmour et al., (1977) used Eq. 3 in the following form:

$$k = [2.303/(t_2 - t_1)] \log C_1/C_2 \quad [4]$$

where k = reaction velocity constant ($time^{-1}$)
t_1 = initial time
t_2 = final time
C_1 = input carbon at t_1
C_2 = residual carbon at t_2

For first order rates of decomposition, the half-life can be calculated from:

$$t_{1/2} = 0.693/k \quad [5]$$

Olson (1963) reviewed simple exponential models in terms of rate of income minus rate of loss for forest litter production. During the same period, Witkamp (1966) in studying decomposition of leaf litter, developed multiple correlations that related temperature, moisture, bacterial counts, and litter age.

The regression formula was given as follows:

$$C = 46.50 + 3.22\,T + 26.86\sqrt{M/D} + 11.39 \log B - 0.64W$$

where C = CO_2 evolution in μliters $hr^{-1}\ g^{-1}$
T = temperature in C
M = moist weight of litter
D = dry weight of litter
B = number of bacterial colonies from 10^{-6} g of air-dry litter
W = number of weeks since leaf fall

The impact of leaf moisture was expressed as $(M/D)^{1/2}$ so as to emphasize influence of low moisture contents. Thus, the latter study stands as one of the first to superimpose the effect of moisture on other decomposition variables. In addition, the regression formula and the weighted values for temperature and moisture or respiration implied a direct proportionality effect. Reuss and Innis (1977) developed a model to assess the effect of water on grassland nitrogen flow using a simulation model. Included was a relationship between soil water suction and coefficients for nitrogen uptake by plants. The rationale included the observation that in most soils, available water is held at a few bars or less (water coefficient near 1.0). In this model, temperature and soil moisture were considered to be multiplicative. Water availability has been shown to be a major variable when evaluating decomposition in a semi-arid grassland (Klein, 1977). In a comprehensive treatment, Hunt (1977) outlined a simulation model for decomposition as part of an ecosystem level model. The relationship between water tension and decomposition rate was based on data published by Bhaumik and Clark (1947). The model functions as a difference equation model, and Hunt (1977) emphasizes that viewing the joint effect of temperature, water, and inorganic nitrogen as the product of independent functions may indeed be an oversimplification. Certainly few data are available to allow for study of pertinent interactions. In a recent review by Paul and VanVeen (1978) reference is made to the development of a computer simulation model to follow the turnover of organic matter in the upper 10 cm of native grassland soil. Soil water content data were obtained as average values from field results and apparently were not treated as a separate parameter. With a more narrow objective in mind, namely to determine the half-life for decomposition of feedlot manure in soil, Gilmour et al., (1977) used a simple numerical expression based on first order rate kinetics as follows:

$$t_{1/2} = \frac{0.693}{k \times (M_2/M_1) \times 0.933^{-[(\text{Annual heat units}/365) - (t_2 - t_1)]}}$$

Field respirometry studies at 60 and 30% of the soil WHC indicated an approximately linear relationship between reaction velocity and moisture content of soil amended with feedlot manure. The following equation was used to estimate the magnitude of the water effect:

$$t_{1/2} = 0.693/[k \times (M_2/M_1)]$$

where k = reaction velocity constant at optimum temperature (27 C)
M_1 = soil water for optimum mineralization
M_2 = actual soil water (field or laboratory)

It was stressed that with the exception of the 60 and 30% WHC moisture values, other k and half-life values were calculated and thereby, may not have been actual representations of the water effect.

DISCUSSION

A considerable amount of research has been conducted to evaluate the influence of water potential on decomposition processes in soils. Unfortunately, numerous studies have reported water content rather than water potential, causing difficulties in comparing results obtained for a variety of soils. In general, water potential appears to influence decomposition of soil organic matter, plant residues, and waste materials by a similar mechanism. There appears to be a two-phase process occurring whereby there is an initial rapid decrease in decomposition within the −0.3 to 10 bar range followed by another region where decomposition decreases linearly with decreasing water potential. As indicated by other studies (Wilson and Griffin, 1975), the role of bacteria is probably minimal once soils attain −15 bar water potentials resulting in actinomycetes and fungi being the major decomposers in soils. This conclusion is consistent with other research where the water potential relations of various actinomycetes and fungi have been evaluated in pure culture systems. The growth rate of these organisms tends to decrease in a linear manner with decreasing water potential. In contrast, bacteria often can metabolize in culture systems at −22 to −100 bars. However, the situation is different in soils since decreasing water potential (decreasing water content) decreases the ability of substrate molecules to diffuse to the bacterial cell and minimizes the ability of bacterial cells to move through the pore space within soils and reach new sources of substrates.

Water potential also influences decomposition processes at high soil water contents where saturated conditions exist resulting in the depletion of O_2 and the development of anaerobic conditions. Anaerobic bacteria are the predominant microorganisms responsible for decomposition in such systems and liberate CO_2 along with partially oxidized C end-products. Methane formation may also be a significant decomposition end-product under some anaerobic conditions.

Concern has been expressed as to the use of water holding capacity as the basis for adjusting soil water content. Historically, soil microbiologists have regarded 50 to 60% of the WHC as the optimum moisture range for microbial growth and metabolic activity. In most instances, the latter moisture range approximates −0.1 to −0.3 bars water potential; and at such water contents, there is little question that biological activity will remain at a high level. The problem relates to interpretation of the water effect in terms of mechanism(s) involved and difficulties encountered when attempting to conduct comparative soil studies. As Griffin (1969) stated, "to absorb water, an organism must be able to overcome the forces tending to retain the water in the soil," and by inference, water uptake becomes energy linked. Certainly, a soil water content based on a percentage of the WHC does not relate directly to thermodynamically available water. However, it may reflect a soil water content which experimentally was found to approach optimum for microbial activity in a particular soil. In this context, the use of WHC has applied value but can hardly be used when attempting to explain the kinetics involved.

At the onset, it would appear that soil water potential has its primary value in being able to provide a common basis for evaluating microbial activity in contrasting soil textural types (sand, loam, clay). Presumably, moisture characteristic curves would then supply comparative data directly applicable to observed microbial activity in soils of varying textural composition. In this regard, Bhaumik and Clark (1947) noted that the moisture tension at which the greatest cumulative amount of carbon dioxide was evolved differed for the five soils studied while maximum CO_2 evolution rates were observed at or near 50 cm tension (with the exception of sand). In studies on plant residue decomposition, Bartholomew and Norman, (1946) noted that decreased microbiological activity could not be entirely attributable to water stress. More recently, various studies (Clark and Kemper, 1967; Cook and Papendick, 1972; Adebayo and Harris, 1971; and Griffin, 1972) have reviewed optimum and minimum water potentials for selected fungal species and related plant diseases.

Studies (unpublished) conducted at the Univ. of Idaho evaluated the predicted half-life values for sludge decomposition in soils maintained at various water contents and associated water potentials. Even though the data presented in Table 4 should be regarded as calculated rather than actual decomposition rates, it is interesting to note that when comparing sludge organic C half-life values in the Palouse silt loam and Ritzville loam, a consistent influence of water availability is not evident. For example, similar half-life values (e.g., 688 and 743 days) were obtained at water potentials of -1 and -15 bars, in the RL and PSL soils, respectively. Thus, essentially the same half-life was obtained at quite different water potentials. Presumptive as these data might be in terms of other soil variables, it remains clear that water potential alone cannot explain the observed differences in half-life value.

SUMMARY

Continued emphasis should be placed upon the utilization of water potential as the characteristic unit for specifying water relations in soils used in decomposition studies. To compare a variety of soils and the relative influence of soil water status, it is imperative that energy units such as water potential be employed. In many cases, processes are also related directly to the water content in a system. For example, a linear relationship between relative N mineralization and water content has been obtained (Stanford and Epstein, 1974), which results in a logarithmic relationship to water potential.

In contrast, summarizing the data available in the literature suggested that relative rates of decomposition were linearly related to the logarithm of water potential. Additional research is needed to confirm the trends observed in Fig. 4 and 7, especially from the standpoint of developing valid simulation models. Furthermore, the impact of excess as opposed to insufficient water on microbial metabolism necessitates a careful choice of methodology to monitor decomposition. For example, CO_2 collection and measurement may underestimate microbial activity in soil approaching water saturation because CH_4 and other reduced C compounds become significant end-products of decomposition under reducing conditions.

Pesticide degradation pathways may be markedly influenced by water potential. This occurs directly through effects on the type and metabolic activity of the soil microflora and indirectly through effects on O_2 concentration. Thus, there is a genuine need to incorporate soil water potential measurements into studies of pesticide degradation in soil to a) allow comparison of investigations in soils of different properties, b) elucidate the direct and secondary effects of water potential on pesticide fate and behavior in soil, and c) subsequently allow the development of soil management practices which enhance the efficiency and effectiveness of pesticides but which minimize potential hazards from their residues.

It is suggested that even if experimental results are presented on the basis of water content that the appropriate mathematical expression be included for conversion of the data to water potential. It would allow for the comparison of data obtained in different investigations. This can be readily done by determining the moisture retention curve for the soil and presenting the mathematical constants in order to calculate water potential from the following equation.

$$\psi = a\,\theta^{-b}$$

where ψ = water potential, − bars
a,b = constants
θ = gravimetric water content, g H_2O/g soil

The problem of units is particularly significant in regard to use of an appropriate index for expression and correction of rate constants derived to characterize the magnitude of the effect of water on rates of decomposition. It would appear at this point that additional study will be required before a satisfactory solution is found. Meanwhile, all three indices, percent water holding capacity, water content, and water potential can each serve a specific experimental objective. From the standpoint of decomposition occurring in aerobic vs. anaerobic environments, the percentage of water filled pores is also a most useful parameter to characterize the soil water status.

LITERATURE CITED

1. Adebayo, A. A., and R. F. Harris. 1971. Fungal growth responses to osmotic as compared to matric water potentials. Soil Sci. Soc. Am. Proc. 35:465–469.
2. Alexander, M. 1967. Biodegradation: problems of molecular recalcitrance and microbial fallibility. Adv. Appl. Microbiol. 7:35.
3. ————. 1968. Degradation of pesticides by soil bacteria. *In* T. R. G. Gray and D. Parkinson (ed.) The ecology of soil bacteria. Univ. of Toronto Press.
4. Bartholomew, W. V., and A. G. Norman. 1946. The threshold moisture content for active decomposition of some mature plant materials. Soil Sci. Soc. Am. Proc. 11:270–279.
5. Bhaumik, H. D., and F. E. Clark. 1947. Soil moisture tension and microbiological activity. Soil Sci. Soc. Am. Proc. 12:234–235.
6. Brooks, G. T. 1974. Chlorinated insecticides. II. Biological and Environmental Aspects. p. 72. CRC Press, Inc., Cleveland, Ohio.
7. Clark, Francis E., and W. D. Kemper. 1967. Microbial activity in relation to soil water and soil aeration. p. 474–480. *In* R. M. Hagen, H. R. Haise, and T. W. Edminister (ed.) Irrigation of agricultural lands. Agronomy 11. Am. Soc. of Agron., Madison, Wis.
8. Cook, R. J., and R. I. Papendick. 1972. Influence of water potentials of soils and plants on root disease. Annu. Rev. Phytopathol. 10:349–374.
9. Edwards, C. A. 1972. Insecticides. p. 513–557. *In* C. A. I. Goring and J. W. Hamaker (ed.) Organic chemicals in the soil environment, Vol. 1, Marcel Dekker, Inc., New York.
10. Fuhr, F., and D. Sauerbeck. 1968. Decomposition of wheat straw in the field as influenced by cropping and rotation. p. 241–250. *In* Soil organic matter studies. Unipub., Inc., New York.
11. Gilmour, C. M., F. E. Broadbent, and S. M. Beck. 1977. Recycling of carbon and nitrogen through land disposal of various wastes. p. 173–194. *In* L. F. Elliott and F. J. Stevenson (ed.) Soils for management of organic wastes and waste waters. Am. Soc. of Agron., Madison, Wis.

12. Greenland, D. J., and P. H. Nye. 1959. Increases in the carbon and nitrogen contents of tropical soils under natural fallows. J. Soil Sci. 10:284–299.

13. Griffin, D. M. 1969. Soil water in the ecology of fungi. Annu. Rev. Phytopathol. 7: 289–310.

14. ————. 1972. Ecology of soil fungi. p. 106–107. Syracuse Univ. Press. Syracuse, N.Y.

15. Guenzi, W. D., and W. E. Beard. 1968. Anaerobic conversion of DDT to DDD and aerobic stability of DDT in soil. Soil Sci. Soc. Am. Proc. 32:522–524.

16. Guenzi, W. D., and W. E. Beard. 1976. The effects of temperature and soil water on conversion of DDT to DDE in soil. J. Environ. Qual. 5:243–246.

17. Hamaker, J. W. 1972. Decomposition: quantitative aspects. p. 255–334. *In* C. A. I. Goring, and J. W. Hamaker (ed.) Organic chemicals in the soil environment, Vol. 2. Marcel Dekker, Inc., New York.

18. Hunt, H. 1977. A simulation model for decomposition in grasslands. Ecology 58:469–484.

19. Jenkinson, D. S. 1965. Studies on the decomposition of plant material in soil. I. Loss of carbon from ^{14}C-labeled rye grass incubated with the soil in the field. J. Soil Sci. 16: 104–115.

20. Jenny, H., S. P. Gessel, and F. T. Bingham. 1949. Comparative study of decomposition rates of organic matter in temperate and tropical regions. Soil Sci. 68:419–432.

21. Johnson, D. D., and W. D. Guenzi. 1963. Influence of salts on ammonium oxidation and carbon dioxide evolution from soil. Soil Sci. Soc. Am. Proc. 27:663–665.

22. Kearney, P. C., and D. D. Kaufman. 1971. Microbial degradation of some chlorinated pesticides. p. 166–189. *In* Degradation of synthetic organic molecules in the biosphere, National Academy of Sciences, Washington, D.C.

23. Klein, D. A. 1977. Seasonal carbon flow and decomposer parameter relationships in a semi-arid grassland soil. Ecology 58:184–190.

24. Laura, R. D. 1974. Effects of neutral salts on carbon and nitrogen mineralization of organic matter in soil. Plant Soil 41:113–127.

25. ————. 1977. Salinity and nitrogen mineralization in soil. Soil Biol. Biochem. 9:333–336.

26. Meikle, R. W. 1972. Decomposition: qualitative relationships. p. 145–225. *In* C. A. I. Goring and J. W. Hamaker (ed.) Organic chemicals in the soil environment, Vol. 2. Marcel Dekker, Inc., New York.

27. Miller, R. H. 1974. Factors affecting decomposition of anaerobically digested sewage sludge in soils. J. Environ. Qual. 3:376–380.

28. Miller, R. D., and D. D. Johnson. 1964. The effect of moisture tension on carbon dioxide evolution, nitrification, and nitrogen mineralization. Soil Sci. Soc. Am. Proc. 28:644–647.

29. Nyhan, J. W. 1975. Decomposition of carbon-14 labeled plant materials in a grassland under field conditions. Soil Sci. Soc. Am. Proc. 39:643–648.

30. ————. 1976. Influence of soil temperature and water tension on the decomposition rate of carbon-14 labeled herbage. Soil Sci. 121:288–293.

31. Oberlander, H. E., and K. Roth. 1968. Transformation of ^{14}C label plant material in soils under field conditions. p. 251–264. *In* Isotopes and radiation in soil organic matter studies. Unipub., Inc., New York.

32. Olson, J. 1963. Energy storage and balance of producers and decomposers in ecological systems. Ecology 44:322–331.

33. Pal, D., and F. E. Broadbent. 1975. Influence of moisture on rice straw decomposition in soils. Soil Sci. Soc. Am. J. 39:59–63.

34. Parr, J. F., and S. Smith. 1974. Degradation of DDT in an everglades muck as affected by time, ferrous iron, and anaerobiosis. Soil Sci. 118:45–52.

35. Paul, E. A., and J. A. VanVeen. 1978. The use of tracers to determine the dynamic nature of organic matter. p. 61–102. Vol. 3. Symposium Session Papers, 11th Int. Congr. Soil Sci., Univ. of Alberta, Edmonton, Canada.

36. Plimmer, T. R., P. C. Kearney, and D. W. Von Endt. 1968. Mechanism of conversion of DDT to DDD by *Aerobacter aerogenes*. J.Agric. Food Chem. 16:594–597.

37. Reddy, K. R., and W. H. Patrick, Jr. 1975. Effect of alternate aerobic and anaerobic conditions on redox potential, organic matter decomposition, and nitrogen loss in a flooded soil. Soil Biol. Biochem. 7:87–94.

38. Reichman, G. A., D. L. Grunes, and F. G. Viets, Jr. 1966. Effect of soil moisture on ammonification and nitrification in the Northern Plain soils. Soil Sci. Soc. Am. Proc. 30: 363–366.

39. Reuss, J. O., and G. S. Innis. 1977. A grassland nitrogen flow simulation model. Ecology 58:374–388.

40. Robinson, J. E. D. 1957. The critical relationship between soil moisture content in the region of wilting point and the mineralization of natural soil nitrogen. J. Agric. Sci. 49:100–105.

41. Sain, P., and F. E. Broadbent. 1977. Decomposition of rice straw in soils as affected by some management factors. J. Environ. Qual. 6:96–100.

42. Shields, J. A., and E. A. Paul. 1973. Decomposition of ^{14}C-labelled plant material under field conditions. Can. J. Soil Sci. 53:297–306.

43. Sindhu, M. A., and A. H. Cornfield. 1967a. Effect of sodium chloride and moisture content on ammonification and nitrification in incubated soil. J. Sci. Food Agric. 18:505–506.

44. ————, and ————. 1967b. Comparative effects of various levels of chlorides and sulfates of sodium, potassium, and magnesium on ammonification and nitrification during incubation of soil. Plant Soil 27:468–471.

45. Singh, B. R., A. S. Argarwal, and Y. Kanehrio. 1969. Effect of chloride salts on ammonium nitrogen release in two Hawaiian soils. Soil Sci. Soc. Am. Proc. 33:557–568.

46. Sommers, L. E., D. W. Nelson, and D. J. Silviera. 1979. Transformations of carbon nitrogen, and metals in soils amended with waste materials. J. Environ. Qual. 8:287–294.

47. Stanford, G., and E. Epstein. 1974. Nitrogen mineralization-water relations in soils. Soil Sci. Soc. Am. Proc. 38:103–107.

48. Stotzky, G. 1972. Activity, ecology, and population dynamics of microorganisms in soil. *In* A. I. Loskin and H. Lechevalier (ed.) Critical reviews in microbiology 2:59–137.

49. Terry, R. E., D. W. Nelson, and L. E. Sommers. 1979. Decomposition of anaerobically-digested sewage sludge as affected by soil environmental conditions. J. Environ. Qual. 8:342–347.

50. Wedemeyer, G. 1967. Dechlorination of 1,1,1-trichloro-2,2-bis(p-chlorophenyl) ethane by *Aerobacter aerogenes*. Appl. Microbiol. 15:569–574.

51. Wetselaar, R. 1968. Soil organic nitrogen mineralization as affected by low soil water potentials. Plant Soil 29:9–17.

52. Wildung, R. E., T. R. Garland, and R. L. Buschbom. The interdependent effects of soil temperature and water content on soil respiration rate in plant root decomposition in arid grassland soils. Soil Biol. Biochem. 7:373–378.

53. Wilson, J. M., and D. M. Griffin. 1975. Water potential and the respiration of microorganisms in the soil. Soil Biol. Biochem. 7:199–204.

54. Witkamp, M. 1966. Decomposition of leaf litter in relation to environment, microflora, and microbial respiration. Ecology 47:194–201.

55. Yoshida, T. 1975. Microbial metabolism of flooded soils. p. 83–122. *In* E. A. Paul and A. D. McLaren (ed.) Soil biochem., Vol. III. Academic Press, New York.

CHAPTER 4

Water Relations in the Life-cycles of Soilborne Plant Pathogens

R. JAMES COOK AND J. M. DUNIWAY[1]

INTRODUCTION

Water relations in the growth, survival, and pathogenicity of plant pathogens is one of the most rapidly expanding subject-areas in plant pathology today. The importance of humidity, dew, rain, and soil moisture to plant disease has been recognized since the beginning of plant pathology, but the real upswing in critical and meaningful studies is only recent and can be traced in part to a paper by D. M. Griffin in 1963. In that paper, Griffin laid some of the groundwork for eventual application of the concept of water potential in the study of the water relations of plant pathogens. Since Griffin's 1963 paper, the subject of water relations in plant disease has been reviewed in no less than 12 review papers or chapters (Cook, 1973, 1980; Cook and Papendick, 1970a, 1972, 1978; Duniway, 1973, 1976b, 1979; Griffin, 1969, 1977; Papendick and Campbell, 1975; Schoeneweiss, 1975), and three books (Baker and Cook, 1974; Griffin, 1972; Kozlowski, 1978). Most of the new information has come from studies of soilborne plant pathogenic fungi.

The topic of water relations in the biology of soilborne plant pathogens has two major aspects: i) the role of water and water potential in the growth, reproduction, and survival of pathogens in the soil and crop residues; and ii) the role of water and water potential in the development of plant disease. The first aspect considers pathogens as soil microorganisms and will make up the greatest part of our chapter. The second aspect considers water in the host-pathogen interaction (Duniway, 1973, 1976b; Kozlowski, 1978; Schoeneweiss, 1975) and is more related to plant physiology than to the subject of this book, which is soil microbiology.

Plant pathogenic fungi persist in soil as dormant propagules such as sclerotia, oospores, thick-walled resting hyphae, and chlamydospores. In

[1]Research plant pathologist, USDA, SEA, AR, Pullman, WA 99164; and associate professor, Dep. of Plant Pathology, Univ. of California, Davis, CA 95616.

Water Potential Relations in Soil Microbiology.

the presence of a suitable host or other source of nutrition these propagules germinate to initiate the new growth of the pathogen. Germination may be direct, i.e., by the growth of hyphae which penetrate the host; or indirect, i.e., the propagule gives rise to sporangia, apothecia, or conidiophores, and from these, zoospores, ascospores, or conidia. These "secondary" propagules can serve as a means of dispersal of pathogens in water or air to the host where they initiate growth and penetrate the host. In the following sections we attempt to summarize how water status, especially when expressed as water potential, plays a significant role in each of these steps leading to infection (Fig. 1) and in the survival of inoculum in soil.

INITIATION AND MAINTENANCE OF HYPHAL GROWTH

Initiation of hyphal growth from dormant propagules is probably a function of the water potential and not the water content of the soil, crop residue, or other medium occupied at the time of germination. Conceivably, germination of large sclerotia are an exception in that like certain plant seeds or roots, the quantity of water required to hydrate the propagules and support germination may exceed the supply, especially in soils where hydraulic conductivity may be limiting. However, the quantity of water needed to support germination of most propagules is very small and probably can be satisfied even by vapor diffusion. Therefore, water potential is probably the more important parameter.

Effects of Water Potential on Spore Germination

The influence of water potential on the direct germination of spores has been studied for conidia of *Verticillium albo-atrum* (Mozumder et al.,

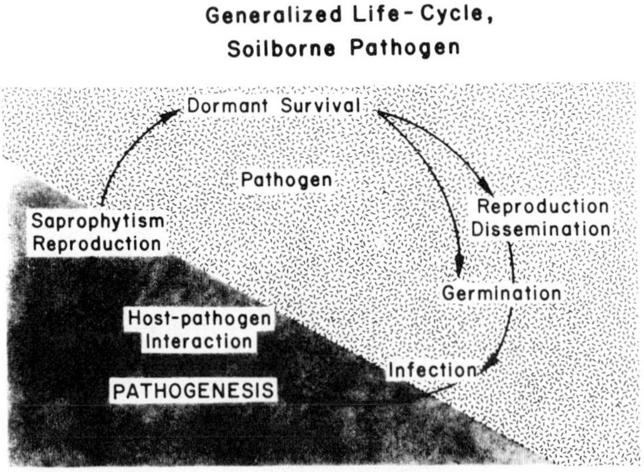

Fig. 1. Generalized life cycle of soilborne plant pathogens. Each step is affected and, in some cases, controlled by water potential of the soil or plant tissue.

1970), *V. dahliae* (Ioannou et al., 1977), *Cephalosporium gramineum* (Bruehl et al., 1972); and *Fusarium roseum* 'Graminearum,' 'Culmorum,' and 'Avenaceum' (Sung and Cook, 1981); for Chlamydospores of *F. roseum* 'Culmorum' (Cook and Papendick, 1970b; Sung and Cook, 1981) and *Phytophthora drechsleri* (Cother and Griffin, 1974); and for ascospores of *Whetzelinia* (= *Sclerotinia*) *sclerotiorum* (Grogan and Abawi, 1975) and *Gibberella zeae* (Sung and Cook, 1981). The studies with *Fusarium* chlamydospores (Cook and Papendick, 1970b) were done in soil at different water contents and hence primarily with variation in the matric component of water potential, whereas the other studies were done with culture media or liquids containing salts or sucrose and hence primarily with variation in the osmotic or solute component of water potential. The three spore types serve entirely different functions in the life cycles of the respective pathogens: the chlamydospores are resistant survival propagules and germinate in soil near host roots; the conidia of the *Cephalosporium* and *Verticillium* wilt pathogens may be produced in the xylem of an infected host and move upward in the transpiration stream, thereby helping systemic invasion of the plant by the pathogen; and the ascospores of *W. sclerotiorum* and *G. zeae* are airborne to aboveground plant surfaces where they germinate and infect.

In all cases, germination occurs at water potentials down to −80 to −90 bars which is approximately the same range of water potential for hyphal growth by these pathogens after germination. Indeed, spore germination in a few cases is even less sensitive to low water potentials than is hyphal growth. Ascospore germination of *W. sclerotiorum* (Grogan and Abawi, 1975) and chlamydospore germination of *F. roseum* 'Culmorum' in soil (Cook and Papendick, 1970b) are apparently unrestricted by water potentials down to −55 or −60 bars, yet subsequent hyphal growth by these two pathogens is maximal at water potentials of −15 bars or above and is progressively more limited with each increment reduction in water potential below about −15 bars (Fig. 2). In *Phytophthora drechsleri*, chlamydospore germination occurs at water potentials down to −98 bars yet subsequent hyphal growth is possible only down to −56 bars (Cother and Griffin, 1974). Increases in water potential can also reduce hyphal growth more than spore germination. For example, macroconidia of *F. roseum* 'Graminearum,' Culmorum,' and 'Avenaceum' and ascospores of *G. zeae* germinate maximally and uniformly well over the range, −1.5 to −20 bars (Sung and Cook, 1981) whereas subsequent hyphal growth is best at about −15 bars and is progressively slower as the water potential is raised from −15 bars up to −1.5 bars (Cook et al., 1972; Cook and Christen, 1976; Wearing and Burgess, 1979). These responses are similar to the germination of ascospores compared with growth of hyphae by *W. sclerotiorum* at osmotic potentials above −15 bars (Fig. 2).

Many spores contain concentrated reserves of food, including lipids, carbohydrates, and proteins. Presumably, the spores with food reserves also have a lower osmotic potential than the more metabolically active hyphae. Thus, hyphal growth would conceivably benefit from an external supply of solutes for osmoregulation in the range between O and −10 to

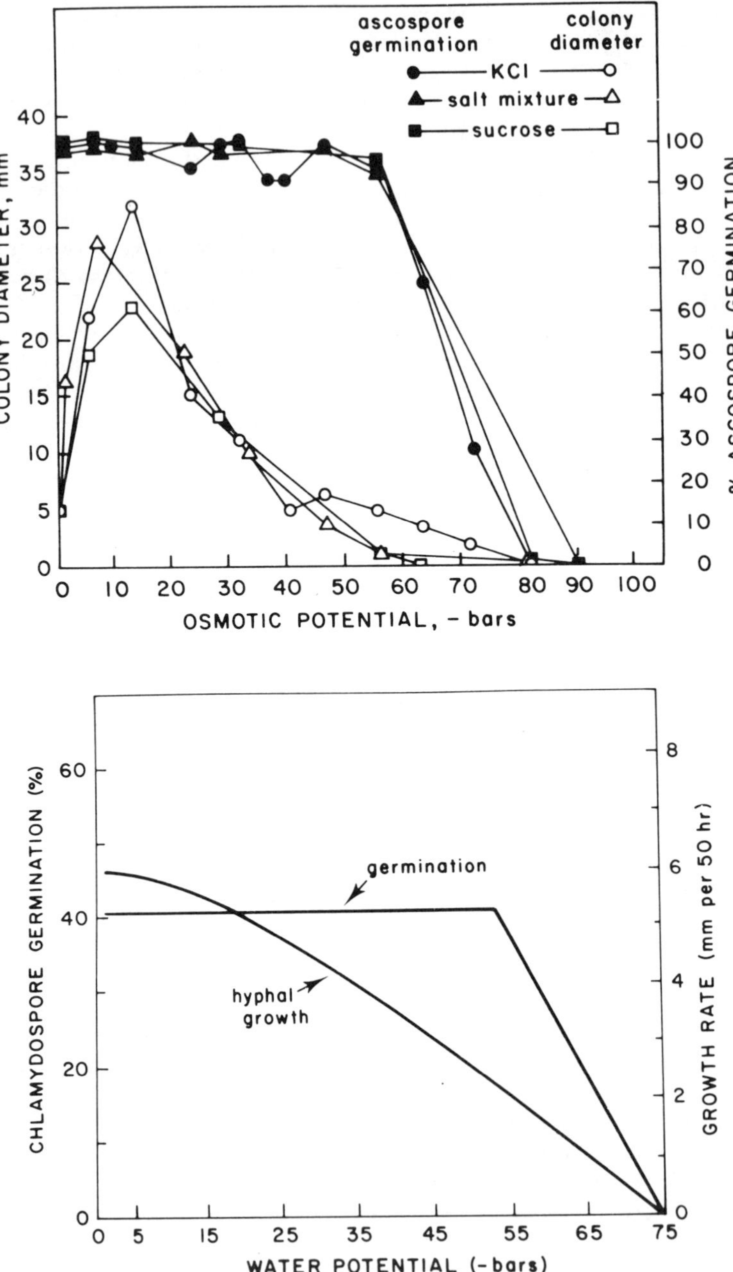

Fig. 2(A, B). Influence of water potential on germination and hyphal growth of two plant pathogenic fungi. (A) Ascospore germination and hyphal growth of *Whetzelinia sclerotiorum* on agar media adjusted to different osmotic potentials with salts or sucrose (Grogan and Abawi, 1975), and (B) Chlamydospore germination (Cook and Papendick, 1970b), and hyphal growth (Cook et al., 1972) of *Fusarium roseum* 'Culmorum' in Palouse silt loam soil at different water (matric) potentials.

−20 bars, whereas spores may be self-sufficient for internal osmotica within this range. This difference could explain why the optimum osmotic potential for hyphal growth is as low as −10 to −20 bars whereas a broader range of osmotic potentials is optimal for spore germination (Fig. 2). Some spores may also achieve a turgor suitable for growth at lower substrate potentials than is possible for hyphae.

The spores of many fungi pathogenic on above-ground plant parts are reported to require free water to germinate. Unfortunately, most experiments with humidity and spore germination have not been designed for accurate control of relative humidity in the range between 96 and 100%, which at equilibrium represents a range in water potential of −60 to 0 bars. Without critical control of air humidity to the first and even second decimal place in the range between 96.00 and 99.99%, the claim that free water (i.e., water at a potential of zero) is needed for spore germination cannot be substantiated. Grogan and Abawi (1975) and Sung and Cook (1981) used osmotically-adjusted agar media to obtain critical control of water potential in the range of −1 to −90 bars and by this method, found that ascospores of *W. sclerotiorum* and *G. zeae*, respectively, could germinate at water potentials down to −85 and −90 bars (equilibrium RH about 94%). Since ascospores are usually thought to require "free water" to infect, it would be interesting to know exactly how much "wettness" they actually require for germination and infection.

Influence of Water on Availability of Nutrients to Pathogen Propagules

Soil water content affects both the amount of plant exudate released into soil (Kerr, 1964; Cook and Flentje, 1967) and the distance exudates move through soil away from the plant surface (Stanghellini and Hancock, 1971). Therefore the availability of nutrients from exudation and not low water potential per se limits germination of propagules of some pathogens in drier soils. This has been demonstrated for oospores of *Pythium aphanidermatum*, for which nutrient availability and not water potential was the limiting factor at −1.0 and −15 bars matric potential (Stanghellini and Burr, 1973). This has also been demonstrated for chlamydospores of *Phytophthora cinnamoni* which germinated at −250 millibars (mb) matric potential near roots of a host, avocado (*Persea indica*), in a clay soil but failed to do so in a sandy soil at matric potentials below −100 mb. Upon addition of nutrients (glucose and asparagine, commonly present in root exudates) to the soil solution, chlamydospores germinated uniformly well in both soils at −250 mb matric potential (Sterne et al., 1977a, 1977b). Similarly, chlamydospore germination by *Fusarium solani* f. sp. *pisi* occurred near pea seeds in soil at 10% but not at 7% water content, even though soil at 7% water content had a water potential well above −15 bars, and was not of itself limiting to chlamydospore germination (Cook and Flentje, 1967). A requirement for external nutrients tends to confine spore germination to the relatively high water potentials that are also suitable for the growth of host plants.

Table 1. Optimal and minimal osmotic potentials for hyphal growth by select pathogenic fungi, as estimated by measuring rates of colony extension on agar media containing salts, sucrose, or both.

Pathogen	Optimum	Minimum	Reference
	bars		
Phycomycetes			
Phytophthora cinnamoni	−10	−40	Adebayo and Harris, 1971
	−15	−45 to −50	Wilson and Griffin, 1975
Phytophthora drechsleri		−56	Cother and Griffin, 1975
Pythium ultimum	−5	−25 to −30	
Ascomycetes and Fungi Imperfecti			
Gaeumannomyces graminis	−1 to −2	−45 to −50	Cook et al., 1972
Fusarium roseum 'Culmorum'	−10 to −15	−90 to −100	Cook et al., 1972
F. roseum 'Graminearium'	−15	−90 to −100	Wearing and Burgess, 1979
F. oxysporum vasinfectum	−10	−115	Manandhar and Bruehl, 1973
Verticillium albo-atrum	−10 to −20	−115	Manandhar and Bruehl, 1973
V. dahliae	−20	−100	Ioannou et al., 1977
Whetzelinia sclerotiorum	−10	−80	Grogan and Abawi, 1975
Fusarium moniliformae	−15	−125 to −140	Wilson and Griffin, 1975
Alternaria tenuis	−10	−100	Adebayo and Harris, 1971
Cercosporella herpotrichoides	−10 to −15	−100	Bruehl and Manandhar, 1972
Cephalosporium gramineum	−2 to −8	−90 to −100	Bruehl et al., 1972
Fusarium nivale	−1 to −3	−45 to −50	Bruehl and Cunfer, 1971
Sclerotina borealis	−10 to −30	−40 to −50	Bruehl and Cunfer, 1971
Basidiomycetes			
Typhula idahoensis	−1 to −2	−30	Bruehl and Cunfer, 1971
Rhizoctonia solani		−45	Dube et al., 1971
Typhula incarnata	−1 to −2	−30	Bruehl and Cunfer, 1971

Optimal and Minimal Water Potentials for Hyphal Growth

The optimal and minimal water potentials for hyphal growth have been studied for a wide array of plant pathogenic fungi (Table 1) simply by measuring growth rate on conventional culture media containing different concentrations of salts, sugars, or both to adjust osmotic potential. In many examples representing Phycomycetes, Ascomycetes, Basidiomycetes, and Fungi Imperfecti, growth is stimulated by reducing the osmotic potential of conventional media (usually about −1 to −2 bars) by −5 to −20 bars. The amount of stimulation and the exact optimal osmotic potential depends on the fungus, and in some cases the solute (Bruehl and Cunfer, 1971), temperature (Manandhar and Bruehl, 1973; Cook and Christen, 1976), or other factors in the environment.

In general, there is an inverse and somewhat linear relationship between fungal growth rate and osmotic potential in the range below the optimal value for growth and extending down to the lower limit for growth of the species. The relationship between colony diameter and osmotic potential for *W. sclerotiorum* (Fig. 2) is, in fact, fairly typical. *Pythium ultimum*, a cause of root rot and damping-off of a wide range of plant species, has one of the highest water potential requirements for growth (lower limit −25 to −30 bars) of all soilborne fungal pathogens tested to date (Table 1). *Phytophthora cinnamoni*, *Gaeumannomyces graminis* var. *tritici*, and *Rhizoctonia solani* (as well as the other Basidiomycetes which have been examined) also have little tolerance of low water potentials, with lower limits approaching −40 to −45 bars (Table 1). In contrast, a variety of plant pathogenic Ascomycetes and Fungi Imperfecti have lower limits for growth in the range of −90 to −120 bars (Table 1). Species of *Aspergillus* and *Penicillium*, which infect fruits and nuts, can grow at water potentials down to −200 bars, and probably lower (Griffin, 1963).

Of the fungi for which optimal and minimal osmotic potentials for growth are known, less than half have also been studied for their ability to grow at various matric potentials. Of those examined, growth has been generally more affected by low matric than by low osmotic potentials. *Gaeumannomyces graminis* var. *tritici* may be an exception in that it exhibits the same lower limit whether exposed to different osmotic or matric potentials (Cook et al., 1972), but *P. cinnamoni* (Adebayo and Harris, 1971), *F. roseum* 'Culmorum' (Cook et al., 1972), *F. oxysporum* f. sp. *vasinfectum* (Manandhar and Bruehl, 1973) and *V. albo-atrum* (Manandhar and Bruehl, 1973) all cease growth at matric potentials −5 to −10 bars higher than their lowest osmotic potentials for growth. Moreover, the stimulation of growth rate by the first reductions in osmotic potential (i.e. the first −5 to −10 bars) has not been duplicated with matric potentials; instead, growth is maximal at matric potentials more nearly zero and is progressively slower with each increment reduction in matric potential down to the lowest value that permits measurable growth (e.g. *F. roseum* 'Culmorum' in Fig. 2).

The ability of some fungi to grow at lower osmotic than matric potentials may result, at least in part, from uptake of the solute by the fungal cell (Griffin in Kozlowski, 1978). Uptake of the solute, e.g., KCl, would lower the internal osmotic potential and help the pathogen maintain turgor and hence continue growth at water potentials where turgor and growth would otherwise be impossible. The stimulation of growth in the range of −5 to −10 bars osmotic potential likewise may result from uptake of the solute, with either or both of two possible benefits: a) lowering of the osmotic potential of the protoplasm to a value more ideal for cellular processes or b) increased turgor and hence acceleration of growth. The response occurs with a wide range of salts as well as sucrose and it appears that fungi may have an array of mechanisms for osmotic adjustment (Duniway, 1979). Presumably, the cell membrane is more permeable to some ions or molecules than to others, and permeability could explain some of the differences in responses of hyphal growth to various solutes (Griffin in Kozlowski, 1978).

Growth Within Host Tissue Related to Water Potential

Much of the active life of plant pathogens occurs within the tissues of a living plant where water potential is controlled more by osmotic than by the kinds of matric forces which can predominate in the bulk soil. For this reason, information on the response of pathogens to various osmotic potentials may predict how a given pathogen might fare in its host under different environmental conditions. Among the pathogens studied so far, several show a close relationship between ability to grow at low osmotic potentials on an agar medium and ability to cause disease of plants growing in a dry environment; or conversely, dependence on high osmotic potentials for growth and the ability to cause disease of plants in wet environments. For example, *G. graminis* var. *tritici* and *F. roseum* 'Culmorum' have lower limits for growth of about −45 and −90 bars osmotic potential, respectively, and cause the most disease under relatively wet and dry soil conditions, respectively. Dryland wheat develops mid-day plant water potentials as low as −30 to −40 bars by the time plants are heading or shortly thereafter, and even lower water potentials at late dough stage (Papendick and Cook, 1974). In contrast, irrigated wheat at the same stages of growth rarely develops water potentials below −20 bars (R. I. Papendick and R. J. Cook, unpublished data). As such, plant water potentials of dryland wheat are at or below the minimum for growth by *G. graminis* var. *tritici* but are ideal for *F. roseum* 'Culmorum.' It should be noted, however, that the growth of most crop plants is confined to much higher water potentials than is the growth of hyphae by most pathogenic fungi (Table 1).

Vascular parasites such as *V. albo-atrum*, *C. gramineum*, and *F. oxysporum* are examples of fungi that grow at very low osmotic potentials, (Bruehl and Cunfer, 1971; Manandhar and Bruehl, 1973; Bruehl et al., 1972; Ioannou et al., 1977), but the diseases they induce are favored by relatively moist soil conditions (Cook and Papendick, 1972). Vascular parasites depend on transpirational flow for carriage of their conidia systemically upward in the xylem of the host (Dimond, 1970). Without ample soil moisture, transpiration—and presumably upward movement of conidia—is reduced. The ability to grow at low water potentials is probably of more value to vascular parasites after plant death, when they must grow from the xylem into the stem parenchyma or other tissues of their dead host and form new propagules to survive in soil. The pathogen will be most successful if this saprophytic colonization process is completed in advance of competition from soil saprophytes which invade once the plant residue is buried. Plants generally die of vascular wilt during the warmest and driest part of the season; an ability to grow at water potentials down to −120 bars, or at 35 C when the water potential is between −35 and −120 bars is added insurance that these fungi will successfully colonize host tissues after plant death.

FORMATION OF REPRODUCTIVE STRUCTURES AND PROPAGULES

Many plant pathogens are dispersed or move to their host as spores formed in reproductive structures in or on soil or crop residues (Fig. 1). Although the number of critical investigations on the water relations of sporulation is still small, and the number in soil even smaller, the process of reproduction by sporulation is clearly very sensitive to soil water potential, soil water content, or both. As the following discussion will show, the water potential requirements for formation of sporangia, apothecia, or other reproductive structures are generally in the range between zero and −15.0 bars and are more exacting than the water potential requirements for initiation and maintenance of growth by the same fungus species. The water potentials needed by pathogens for sporulation are frequently within the range of water potentials needed for growth by higher plants; this may sometimes help to synchronize the activities of soilborne plant pathogens with periods of active growth by their host.

The most complete studies on water potential requirements for sporulation have been conducted with important pathogenic members of the Phycomycetes (species of *Phytophthora*, *Pythium*, and *Aphanomyces*) which cause root rots and seedling blights on many plant species, particularly under very wet or even water-logged soil conditions. The predominate cycle in asexual reproduction begins with formation of a sporangium, within which flagellated zoospores are later formed by cleavage of the cytoplasm. The zoospores are finally discharged from the sporangium and swim actively in the soil solution. Zoospores exhibit chemotaxis towards host tissues and they can be an abundant and very effective form of inoculum (Duniway, 1979).

Each of the steps in sporangium and zoospore formation is highly sensitive to water potential. For example, *Phytophthora cryptogea* (*P. drechsleri*) only forms sporangia in soil at water potentials down to −4 bars (Duniway, 1975a, 1975b). When the matric component of soil water potential was increased to induce the final differentiation and release of zoospores by sporangia of *P. cryptogea* and *P. megasperma*, release was rapid at 0 bars (saturation) and −1 mb (millibar), was impaired at −5 mb, greatly restricted at −10 mb, and did not occur at −25 mb matric potential (MacDonald and Duniway, 1978a). The results on zoospore discharge were the same for both species and in soils of different textures (Fig. 3) suggesting that zoospore release is affected more by matric po-

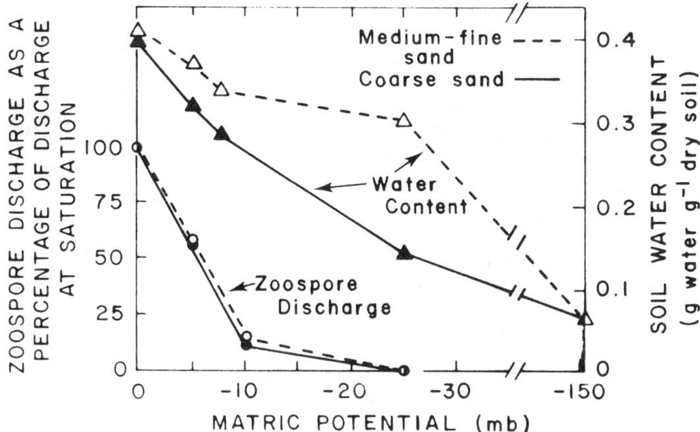

Fig. 3. Influence of matric potential on soil water content and zoospore discharge by *Phytophthora cryptogea* in the coarse (>250 μm, closed symbols) and medium-fine (60 to 250 μm, open symbols) sand fractions of Yolo fine sandy loam. Soil water content and numbers of zoospores were measured 6 hours after matric potential was increased from −150 mb to the higher values shown (MacDonald and Duniway, 1978a).

tential than by soil water content. Pfender et al. (1977) obtained somewhat similar results with *P. megasperma*; zoospore discharge was maximal at 0 mb and was reduced by more than 90% at −50 mb matric potential. A terrestrial chytrid, *Olpidium brassicae*, also requires matric potentials higher than −60 mb for a rapid release of zoospores (Westerlund et al., 1978).

In contrast to the results obtained with soil water under matric control, zoospore release by *P. cryptogea* and *P. megasperma* in solutions of salts, sugars, or polyethlene glycols (PEG) occurs at osmotic potentials as low as −6 to −9 bars (MacDonald and Duniway, 1978a). Only in PEG 6000 was zoospore discharge inhibited at relatively high (−1.3 bars) potentials, but even this value is considerably below the lower limit with soil matric potential. Probably the sporangia of *Phytophthora* absorb the solutes in some solutions and thereby acquire the ability to discharge zoospores at osmotic potentials lower than would be possible at equivalent matric potentials. This suggestion is supported by the fact that zoospore discharge with *P. cryptogea* and *P. megasperma* occurred at even lower osmotic potentials in solutions of PEG 300 than with PEG 6000 (MacDonald and Duniway, 1978a). Similarly with zoospore discharge in *Aphanomyces euteiches*, the lower limit was nearly −3 bars osmotic potential in solutions of glucose, KCl, or other similar solutes, but was only −0.5 bars in a solution of PEG 20,000 (Hoch and Mitchell, 1973). Studies on zoospore discharge and plasmolysis in sporangia of *Phytophthora cactorum* by Gisi et al. (1977) suggest further that effects of specific ions on zoospore discharge can be attributed largely to differential permeability. The results obtained with control of soil matric potential is the range of zero to −25 mb are probably the most relevant to the natural occurrence of zoospore discharge by plant pathogens in soil (Duniway, 1979). Sporangia of *Phytophthora* and *Pythium* species can also germinate directly by the growth of germ tubes; while sporangia are frequently found to shift away from zoospore release towards direct germination with relatively small decreases in matric or osmotic potential (Duniway, 1979), the lower limits of water potential for direct germination are unknown.

Recent experiments have shown that some fungi have even higher water potential requirements for zoospore discharge than do *A. euteiches* and *Phytophthora* species. *Pythium iwayamai* and *P. okanogenense* both released maximum numbers of zoospores at 0 mb matric potential, but released none at −1 mb or lower matric potentials (Lipps, 1980). More zoospores were released when a 1-mm layer of water covered the soil surface than when the soil was only saturated. In solutions, zoospore release by these *Pythium* spp. was limited to osmotic potentials higher than −0.5 bar. A soil extract at −0.25 bar osmotic potential was completely inhibitory to their zoospore release. These fungi rot wheat leaves under snow and ice and release zoospores with greatest frequency in freshly melted snow water; apparently zoospore release in this case is highly sensitive to the concentration of certain ions in the soil solution (Lipps and Bruhel, 1978).

Sclerotia of *Whetzelinia sclerotiorum* germinate in soil by production of several apothecia, each on the apex of a stipe that extends above ground. Ascospores are ejected from the apothecia and are carried by air currents to the surfaces of host leaves. Production of apothecia occurred in water (0 bars) but not in solutions adjusted to −6 bars osmotic potential with KCl, a salt mixture, or sucrose; this is in considerable contrast to the −90 bar lower limit is osmotic potential for mycelial growth by the same fungus (Grogan and Abawi, 1975). Production of apothecia is even more sensitive to soil matric potential, with the number of initials and mature apothecia being maximal at −0.08 to −0.38 bar and the growth of mature apothecia being prevented at matric potentials below about −0.5 bar (Duniway et al., 1977). These effects of soil matric potential were determined with hydrated sclerotia and at nearly saturated ambient humidities. If significant water uptake is required, either for hydration or transpiration, the growth of apothecia in nature might be confined to even higher soil matric potentials, i.e., those where the hydraulic conductivity is sufficient.

Water potential requirements may differ among isolates of a fungal species or subspecies, suggesting the existence of ecotypes adapted to conditions under which they occur. For example, isolates of *Fusarium roseum* 'Graminearum,' 'Culmorum,' and 'Avenaceum' from drier areas of Washington State all formed macroconidia maximally at −15 bars osmotic potential, in contrast to isolates of *F. roseum* 'Graminearum' from Pennsylvania which formed macroconidia maximally at −1.4 bars, the highest water potential tested (Sung and Cook, 1981). None of the isolates produced macroconidia at osmotic potentials below −60 to −80 bars (Sung and Cook, 1981), although this species can grow as mycelium at osmotic potentials down to −90 to −100 bars (Cook and Christen, 1976; Wearing and Burgess, 1979). The Pennsylvania isolates also produced the sexual stage (perithecia, = *Gibbellera zeae*) in contrast to Washington isolates which produced only macroconidia. Production of perithecia by the Pennsylvania isolates responded to water potential more nearly like that of production of macroconidia of the Washington isolates, being maximal at about −15 bars (Sung and Cook, 1981). The results suggest that Pennsylvania isolates are adapted to maximal production of macroconidia under humid conditions and that as host tissues dry, they not only produce fewer macroconidia, but make perithecia instead.

In contrast to reproductive structures, which frequently require the same range of water potentials as higher plants, many propagules formed as survival structures in soil or on crop residues can be formed over a range of water potentials that more nearly resemble the range for hyphal growth. For example, microsclerotial formation by *Verticillium dahliae* was maximal in the range of −20 to −30 bars, but occurred at soil water potentials down to −100 bars (Ioannou et al., 1977). Microsclerotial formation was also inhibited in saturated soil, apparently because of lack of aeration (Ioannou et al., 1977). Chlamydospores and oospores of some *Phytophthora* spp. can evidently be formed at soil water potentials too dry for the formation of sporangia or zoospores (Reeves, 1975; Sneh & McIntosh, 1974).

MOVEMENT OF PATHOGENS IN SOIL

Active movement by zoospores of plant pathogenic fungi to infection sites has been demonstrated many times (Hickman, 1970). There is no doubt that zoospores are effectively dispersed, both actively and passively, in surface water. However, the size of water-filled pores and pore-size distribution, determined by water potential and soil structure, can greatly impede zoospore movement through soil (Griffin, 1969, 1972; Cook and Papendick, 1972). For example, as one might expect from filtration theory, the first experiments dealing with soil constraints on zoospore dispersal found only limited downwards movement of *Phytophthora infestans* to infect potato tubers (Zan, 1962). In fact, Stanghellini (1974) has suggested that soil constraints on movement of *Pythium* zoospores are so great that zoospore movement may not be a major factor in infection by members of this important genus. Some zoospore movement in soil, however, has been demonstrated in a variety of experiments (Duniway, 1976a), and we are just beginning to learn the extents to which soil texture and water status actually limit active zoospore movement. *Phytophthora cryptogea* zoospores were able to swim through a coarse-textured soil at greater than −1 mb matric potential, whereas active movement in the same soil was reduced at a matric potential of −10 mb and was not detected at −50 mb (Duniway, 1976a) (Fig. 4). Active movement in finer-textured loam soils was limited to 5 mm at −1 mb matric potential and was not detected in two of three soils at −10 mb (Duniway, 1976a). These results suggest that textural constraints on zoospore movement are indeed great, but they also demonstrate the levels of moisture which are critical if zoospore movement through soils is to have any role in infection.

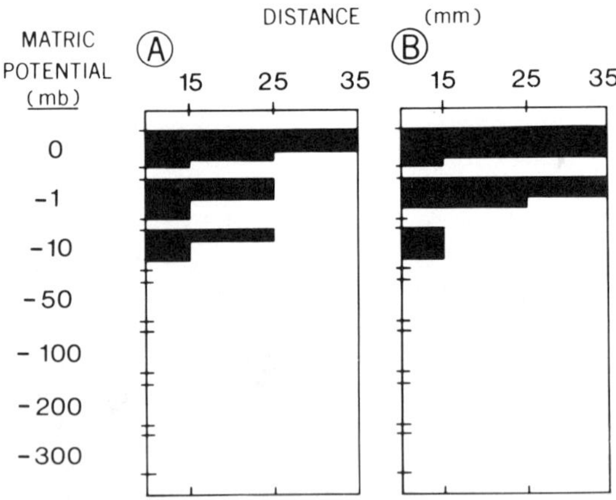

Fig. 4(A, B). Effect of matric potential on the movement of *Phytophthora cryptogea* zoospores to safflower seedling roots at various distances from the point where a zoospore suspension was added to a coarse-textured soil mix. *Phytophthora cryptogea* was detected in seedlings by (A) symptom development and (B) isolation on selective medium. Bar width is proportional to the incidence of *P. cryptogea* in seedlings with the maximum distance between marks on the vertical axis representing 100% incidence (Duniway, 1976a).

Soil pores on the order of 300 μm in diam drain water at −10 mb matric potential (Duniway, 1976a; Griffin, 1972). Therefore, the matric potential requirements for active movement by *P. cryptogea* zoospores suggest that there must not only be continuous water-filled channels that can accommodate zoospores (diam < 60 μm with flagella extended), but probably the larger and less tortuous the water-filled pores the more suitable they will be for active zoospore movement (Duniway, 1976a). In fact, few soils appear to contain a significant volume of macropore space to accommodate swimming zoospores (Cook and Papendick, 1972). The prevalence of such large pores probably depends upon soil texture, aggregation, compaction, and the presence of channels left by dead roots and burrowing animals. Therefore, it is difficult to predict zoospore behavior in many natural soils from the results obtained in sieved and reconstituted soils (Duniway, 1976a). More realistic data are needed on the macropore structure and zoospore movement in undisturbed soils.

There are also a number of biological parameters that affect the extent of active inoculum dispersed in soil. In *Phytophthora* spp. zoospores tend to encyst when they make contact with surfaces and this reduces the duration of their motile period. In addition, zoospores of *P. cryptogea* are better swimmers than those of *P. megasperma* in soil (MacDonald and Duniway, 1978b), and zoospores of *Olpidium brassicae*, which do not commonly encyst, swim still farther in finer soils and at slightly lower matric potentials (Westerlund et al., 1978) than do zoospores of *P. cryptogea* (Duniway, 1976a). Evidently, species that differ in their swimming behavior or motile period may have different capacities for movement through soil. Furthermore, while research on *Phytophthora palmivora* suggests that motile zoospores function more effectively as inoculum in natural soils than do nonmotile spores (Kliejunas and Ko, 1974), in most instances we have no firm estimates of the extent to which active zoospore movement enhances root infection over that which could occur by other means such as the growth of germ tubes.

There are also important interactions between soil texture, water content or matrix potential, and the active movement of plant pathogenic bacteria and nematodes (Griffin, 1972; Jones, 1975). However, because bacteria are very much smaller than fungal zoospores and because nematodes have a very different means of locomotion, their water requirements for movement through soil are not so stringent as are the water requirements for motile zoospores. In fact, nematode movement is highly restricted in saturated soils and is largely confined to the −20 to −200 mb matric potential range (Jones, 1975).

SURVIVAL AND LONGEVITY OF PATHOGENS IN SOIL

Water and water potential have both direct and indirect effects on the survival and longevity of pathogens in soil. The direct effects include death through desiccation at the one extreme, and death with flooding and hence lack of oxygen at the other extreme. Indirect effects are mediated through the antagonistic effects of the associated microbiota.

Most soilborne plant pathogens are equipped during dormancy to withstand considerable desiccation. Chlamydospores, oospores, and thick-walled hyphae and conidia are particularly insensitive to long periods of low water potential. Indeed many pathogens actually survive longer in dry than in wet soil because of slower basal metabolism of the propagule, less activity of the associated microbiota, or both. *Cochliobolus sativus*, cause of common root rot of wheat and barley (*Hordeum vulgare* L.), exists in soil as thick-walled, pigmented, multicellular conidia that are virtually unaffected by drying. In the laboratory, the conidia remain 90 to 95% viable for years when stored in talcum powder at 4 C (Timian, 1959); and in soil, 100% of the conidia survived for 52 months at an equilibrium relative humidity (RH) of 50% or less, but lost viability during the same time period at 70 to 100% RH (Ledingham, 1970). Likewise, *Gaeumannomyces graminis* var. *tritici* survived 46 weeks as hyphae in straw in soil at −950 bars, but quickly lost viability, or was displaced in the straw by soil saprophytes, when soil water potentials were near −1 bar (MacNish, 1973). Air-drying may even help break a constitutive dormancy in *Pythium oospores* (Lumsden and Ayres, 1975).

Among the propagules used for dispersal by soilborne plant pathogens, those which move in water are usually very sensitive to desiccation whereas those that move in air are not. Ascospores of *Whetzelinia sclerotiorum* are an example of airborne propagules that are resistant to drying. When exposed to different relative humidities, they lost their ability to germinate after only 36 hours at 100% RH, while the same loss took 264 hours at 65%, and 504 hours at 45%; nearly all of the ascospores remained completely germinable after 21 days (duration of the test) at 10% RH (Grogan and Abawi, 1975). In contrast, most zoospores of *Phytophthora megasperma* were killed within 1 week in soil dried to only −20 bars water potential and those of *P. cinnamomi* remained viable only at soil water potentials higher than −42 bars (MacDonald and Duniway, 1979).

Propagules with thick walls are generally more resistant to desiccation than spores with thin walls. The microsclerotia (survival structures) of *Verticillium dahliae* are resistant to desiccation but the microconidia (thin-walled dissemination structures) of the same fungus are not. A common practice in estimating the population of this pathogen in soil is to air-dry the soil to eliminate the microconidia and thereby permit an estimate mainly of the microsclerotia (Menzies and Griebel, 1967). Oospores of *Pythium ultimum* are resistant to drying while thick-walled but become sensitive to drying as their walls become thin (Lumsden and Aryes, 1975). On the other hand, the walls of chlamydospores of *Fusarium roseum* 'Graminearum' viewed by transmission electron microscopy appear no different from those of the closely related *F. roseum* 'Culmorum,' yet chlamydospores of Culmorum survive rapid drying but those of Graminearum do not (Sitton and Cook, 1980). Wall structure alone cannot explain resistance of all spores to drying. Probably amounts of constitutive osmotica and other physiological characters are also important.

Sclerotia exhibit wide variations in resistance to desiccation, depending on the fungus species and other factors. Sclerotia of *Phymatotrichum omnivorum* are among the most sensitive to drying (King et al., 1931). The fungus is a native inhabitant of the arid Southwest and Mexico but rarely persists for more than a few weeks in a dry tillage layer. Instead, the fungus survives and infects the roots of its hosts at the lower depths, sometimes as deep as 4 m. Sclerotia of *Sclerotium oryzae* lose a substantial amount of their weight and lose viability during repeated cycles of wetting and drying (Keim and Webster, 1974). According to Menzies and Griebel (1967) microsclerotia of *Verticillium dahliae* become more sensitive to drying after supporting sporulation (a form of germination whereby conidiophores and conidia form on the microsclerotia). With *Sclerotium rolfsii*, drying then soaking sclerotia formed on culture media triggers spontaneous germination (Smith, 1972), whereas sclerotia formed in field soil are slow to germinate whether dried before soaking or not and require a volatile from peanut hay to trigger germination (Beute and Rodriguez-Kabana, 1979).

With some soilborne propagules, additional time is needed for germination to occur after exposure to drying. Baker and Cook (1974) have termed this effect "desiccation dormancy." For example, chlamydospores of *Fusarium solani* f. sp. *phaseoli*, which normally require about 4 hours to germinate under conditions of favorable environment and nutrition, require about 40 hours to germinate once exposed for 3 or more months to very dry soil conditions. Linderman and Toussoun (1968) observed a similar effect of drying on chlamydospores of *Thielaviopsis basicola*. The additional time required for germination is apparently needed for the chlamydospore to rehydrate, and perhaps to recover from other adverse effects of desiccation (Cook and Papendick, 1972). A fungus that requires 40 hours to germinate rather than only 4 hours may also be at a competitive disadvantage in soil.

The success or failure of pathogen interactions with antagonistic microorganisms is often influenced by soil water potential. This is illustrated by *Fusarium roseum* 'Culmorum' which escapes the antagonistic effects of soil bacteria by virtue of its ability to grow at soil water potentials below the limit for growth of most soil bacteria (Cook and Papendick, 1970b). The germ tubes from chlamydospores of this fungus have been observed to elongate and branch in soil for up to 6 to 7 days at soil water potentials below about −15 bars. In contrast, the germ tubes lyse or the fungus forms new chlamydospores within 24 to 48 hours after germination if soil water potentials are between 0 and −15 bars where bacteria are active. When segments of wheat straw were buried for 1 week in nonsterile soil containing a high population of Culmorum, the fungus, acting as a saprophyte, colonized the greatest proportion of the straws at soil water potentials of −80 to −90 bars (Cook, 1970). Even though these water potentials are near the lower limit for growth of Culmorum (Cook et al., 1972), the niche presented at −80 to −90 bars was apparently more available to this organism than the niche presented at higher potentials.

Giant vampyrellid amoebae (*Arachnula impatiens* Cienk.) are part of the microbiotia antagonistic to fungi in wet soils, but are also readily escaped by fungi if the soil dries even to field capacity. These amoebae make holes (1–2 μm in diam) in the pigmented walls of a wide array of fungal spores (Old, 1977; Anderson and Patrick, 1978) and also the runner hyphae of the take-all fungus, *G. graminis* var. *tritici* (Homma et al., 1979). Once a hole is formed, cell contents spill out and the cell dies. The amoebae are inactive in saturated soil (0 bars), apparently because of inadequate O_2 (Cook and Homma, 1979). Maximal activity occurs at about −25 mb matric potential in sandy soils, and at −100 to −150 mb in loam or clay soils, i.e., at matric potentials which permit air to enter the soil. The amoebae are again inactive in the soil at about −200 mb matric potential and in solution culture at about −400 to −500 mb osmotic potential (Cook and Homma, 1979). Amoebae may play a role in the long-term degradation of resting spores and hyphae of fungi in soil, but owing to their high water potential requirements, are probably of little importance against plant pathogens during active parasitism of plants in soils at field capacity or drier.

Some of the most important antagonistic interactions with soil water potential occur during dormant survival of pathogens. This is especially true for pathogens dependent on crop residues as a source of food and as a place to survive and reproduce. For example, *Cephalosporium gramineum*, cause of a vascular disease of wheat, produces an antibiotic during its saprophytic existence in wheat straw in soil that discourages other would-be colonists of the straw. However, the fungus does not produce antibiotic below −75 to −100 bars water potential, and consequently is over-run in the straw by fungi able to grow at these potentials, particularly *Penicillum*, *Aspergillus*, and *Fusarium* species (Bruehl and Lai, 1968). Nevertheless, the fungus is virtually the sole possessor of the straw at still lower water potentials (below −270 bars) because at these potentials even the competitors are inactive.

SUMMARY AND CONCLUSIONS

Germination, growth, and reproduction by soilborne plant pathogens each is affected and even controlled by the water potential of the site (niche) occupied by the pathogen, whether soil, rhizosphere, living host tissue, or dead crop residue. Indeed, plant pathogens not only function successfully within the range of water potentials prevalent in soil or plant tissues during crop growth, some aspects of their water relations may even tend to synchronize their activities with the growth of a host.

The numerous plant pathogenic fungi which have been studied to date all grow as mycelium at osmotic potentials down to −25 to −30 bars, and many down to −100 to −120 bars. These osmotic potentials are well below those of living plant tissues where much or most pathogen growth ocurs. Most pathogens make maximal growth in the range of −5 to −15 bars osmotic potential which prevails in crop plants. In the few cases compared, matric potentials have been found to restrict hyphal growth more than equivalent low osmotic potentials.

In general, pathogens that only cause disease of plants under moist soil conditions, and thus where plant water potentials are high, are incapable of growth at osmotic potentials below about −30 to −40 bars. Conversely, pathogens that cause disease of plants under dry soil conditions, and thus where plant water potentials are low, are capable of growth at osmotic potentials down to −100 to −120 bars. Among the important exceptions are the vascular wilt fungi (*Fusarium oxysporium*, *Verticillium albo-atrum*, *V. dahliae*, and *Cephalosporium gramineum*) which grow at osmotic potentials down to −100 to −120 bars but which incite the most disease under conditions of ample soil moisture for plant growth. The ability of the vascular wilt fungi to grow at low osmotic potentials probably is an advantage for colonization and reproduction in wilted or dead host tissues before the tissues become colonized by saprophytes.

For fungal pathogens studies to date, the initiation of hyphal growth by direct germination of spores or other propagules require essentially the same range of water potentials as does continuing growth of mycelium. However, the availability of nutrients in soil and not water potential per se may be the limiting factor to germination and infection. Nutrients needed by a pathogen often must diffuse from a root, seed, or other plant part as exudates and in these cases, the volume of water as films on soil particles and in pores may govern nutrient diffusion to a propagule. By this mechanism, germination by pathogens otherwise capable of growth at very low potentials may be confined to the higher range of water potentials suitable for plant growth.

The limited studies to date indicate that some of the reproductive stages in many fungi are more sensitive to reduced water potential than are spore germination and hyphal growth. Examples include apothecium formation by *Whetzelinia sclerotiorum* and perithecium formation by *Gibberella zeae*; the former is limited to greater than −6 bars osmotic potential (higher still if under matric control), the latter to −40 to −50 bars osmotic potential, even though ascospores of both fungi germinate and hyphae grow at osmotic potentials down to −80 to −100 bars. Plant pathogens with motile zoospores that contact and infect their hosts (e.g. species of *Phytophthora*, *Pythium*, and *Aphanomyces*), although apparently capable of hyphal growth at osmotic potentials as low as −25 to −30 bars, require much higher water potentials to form zoosporangia, and still higher potentials to form and release zoospores. Indeed, some members of this group release zoospores only at 0 mb matric potential, others only down to −25 mb matric potential. This sensitivity of zoosporogenesis to matric potential is advantageous in that zoospores are not formed when soil is too dry for their active or passive dispersal. In contrast to reproductive stages, soilborne plant pathogens can apparently form resting propagules (e.g. microsclerotia or chlamydospores) within roughly the full range of water potentials that permit hyphal growth.

Studies of the influence of soil water on plant disease have been carried out over the past 50 years or more, but critical studies on how water affects plant disease are considerably more recent and generally had to await development of the water potential concept and instrumentation to measure water potential and its components. A great deal of valuable information has accumulated rapidly, simply by determining optimal and minimal matric and osmotic potentials for hyphal growth. Equally valuable has been the concurrent and rapid accumulation of information on water potentials of tissues of host plants growing under various conditions. We now realize that the living host tissues in which pathogens grow in nature are almost universally several bars lower in water potential (drier) than are the media normally used to culture pathogens in the laboratory.

Work on water potential requirements for reproduction of plant pathogens, although in very preliminary stages, already shows excellent promise as a basis for revealing additional ways that water potential affects plant disease. However, almost no work has been done to date on water relations at the cellular or metabolic level for pathogens or for the host-pathogen interactions. Critical studies are also needed to reveal why microorganisms commonly respond differently to the osmotic and matric components of water potential, why different pathogens have different optimal and minimal water potentials for growth, or why different stages in the life cycle of a given pathogen have different optimal and/or minimal water potentials. Work is also needed to reveal how water potential affects the formation of enzymes, toxins, elicitors, or other important molecules in pathogenesis and host resistance mechanisms. Hopefully our paper will bring an awareness to students and established researchers of the importance of water potential in the biology of plant pathogens, the enormous opportunities available for meaningful and creative research in this area, and the potential for pay-off in plant disease control.

LITERATURE CITED

1. Adebayo, A. A., and R. F. Harris. 1971. Fungal growth responses to osmotic as compared to matric water potential. Soil Sci. Soc. Am. Proc. 35:465–469.
2. Anderson, T. R., and Z. A. Patrick. 1978. Mycophagus amoeboid organisms from soil that perforate spores of *Thielaviopsis basicola* and *Cochliobolus sativus*. Phytopathology 68:1618–1626.
3. Baker, K. F., and R. J. Cook. 1974. Biological control of plant pathogens. W. H. Freeman and Co., San Francisco.
4. Beute, M. K., and R. Rodriguez-Kabana. 1979. Effect of wetting and the presence of peanut tissues on germination of sclerotia of *Sclerotium rolfsii* produced in soil. Phythopathology 69:869–872.
5. Bruehl, G. W., and B. Cunfer. 1971. Physiologic and environmental factors that affect the severity of snow mold of wheat. Phytopathology 61:792–799.
6. ————, ————, and M. Toiviainen. 1972. Influence of water potential on growth, antibiotic production, and survival of *Cephalosporium gramineum*. Can. J. Plant Sci. 52:417–423.
7. ————, and P. Lai. 1968. The probable significance of saprophytic colonization of wheat straw in the field by *Cephalosporium aramineum*. Phytopathology 58:464–466.
8. ————, and J. B. Manandhar. 1972. Some water relations of *Cercosporella herpotrichoides*. Plant Dis. Rep. 56:594–596.
9. Cook, R. J. 1970. Factors affecting saporphytic colonization of wheat straw by *Fusarium roseum* f. sp. *cerealis* 'Culmorum.' Phytopathology 60:1672–1676.
10. ————. 1973. Influence of low plant and soil water potentials on diseases caused by soilborne fungi. Phythopathology 63:451–458.

11. ———. 1980. Water relations in the biology of *Fusarium*. *In* P. E. Nelson, T. A. Tousoun, and R. J. Cook (ed.) *Fusarium*: Biology and taxonomy. Pennsylvania State Univ. Press, University Park.

12. ———, and A. A. Christen. 1976. Growth of cereal root-rot fungi as affected by temperature-water potential interactions. Phytopathology 66:193–197.

13. ———, and N. T. Flentje. 1967. Chlamydospore germination and germling survival of *Fusarium solani* f. *pisi* in soil as affected by soil water and pea seed exudation. Phytopathology 57:178–182.

14. ———, and R. I. Papendick. 1970a. Effect of soil water on microbial growth, antagonism, and nutrient availability in relation to soil-borne fungal diseases of plants. p. 81–88. *In* T. A. Toussoun, R. V. Bega, and P. E. Nelson (ed.) Root diseases and soil-borne pathogens. Univ. of California Press, Berkeley.

15. ———, and ———. 1970b. Soil water potential as a factor in the ecology of *Fusarium roseum* f. sp. *cerealis* 'Culmorum.' Plant Soil 32:131–145.

16. ———, and ———. 1972. Influence of water potential of soils and plants on root disease. Annu. Rev. Phytopathol. 10:349–374.

17. ———, and ———. 1978. Role of water potential in microbial growth and development of plant disease with special reference to postharvest pathology. HortScience 13: 559–564.

18. ———, ———, and D. M. Griffin. 1972. Growth of two root-rot fungi as affected by osmotic and matric water potentials. Soil Sci. Soc. Am. Proc. 36:78–82.

19. Cother, E. J., and D. M. Griffin. 1974. Chlamydospore germination in *Phytophthora drechsleri*. Trans. Br. Mycol. Soc. 63:273–279.

20. Dimond, A. E. 1970. Biophysics and biochemistry of the vascular wilt syndrome. Annu. Rev. Phytopathol. 8:301–322.

21. Dube, A. J., R. L. Dodman, and N. T. Flentje. 1971. The influence of water activity on the growth of *Rhizoctonia solani*. Austr. J. Biol. Sci. 24:57–65.

22. Duniway, J. M. 1973. Pathogen-induced changes in host water relations. Phytopathology 63:458–466.

23. ———. 1975a. Limiting influence of low water potential on the formation of sporangia by *Phytophthora drechsleri* in soil. Phytopathology 65:1089–1093.

24. ———. 1975b. Formation of sporangia by *Phytopahthora drechsleri* in soil at high matric potentials. Can. J. Bot. 53:1270–1275.

25. ———. 1967a. Movement of zoospores of *Phytophthora cryptogea* in soils of various textures and matric potentials. Phytopathology 66:877–882.

26. ———. 1976b. Water status and imbalance. p. 430–449. *In* R. Heitefuss and P. H. Williams (ed.) Encyclopedia of plant physiology. New Series Vol. 4, Physiological Plant Pathology. Springer-Verlag, Berin.

27. ———. 1979. Water relations of water molds. Annu. Rev. Phytopathol. 17:431–460.

28. ———, G. S. Abawi, and J. R. Steadman. 1977. Influence of soil moisture on the production of apothecia by sclerotia of *Whetzelinia sclerotiorum*. Proc. Am. Phytopathol. Soc. 4:115.

29. Gisi, U., J. J. Oertli, and F. J. Schwinn. 1977. Wasser- und Salzbeziehungen der Sporangien von *Phytophthora cactorum* (Leb. et Cohn) Schroet, in vitro. Phytopathol. Z. 89:261–284.

30. Griffin, D. M. 1963. Soil moisture and the ecology of soil fungi. Biol. Rev. 38:141–166.

31. ———. 1969. Soil water in the ecology of fungi. Annu. Rev. Phytopathol. 7:289–310.

32. ———. 1972. Ecology of soil fungi. Chapman and Hall. London.

33. ———. 1977. Water potential and wood-decay fungi. Annu. Rev. Phytopathol. 15: 319–329.

34. ———. 1980. Water potential as a selective factor in microbial studies. p. 00–00. *In* J. F. Parr and W. R. Gardner (ed.) The concept of water potential applied to soil microbiology and biochemistry. Am. Soc. Agron., Madison, Wis.

35. Grogan, R. G., and G. S. Abawi. 1975. Influence of water potential on growth and survival of *Whetzelinia sclerotiorum*. Phytopathology 65:122–138.

36. Hickman, C. J. 1970. Biology of *Phytophthora* zoospores. Phytopathology 60:1128–1135.
37. Hoch, H. C., and J. E. Mitchell. 1973. The effects of osmotic water potentials on *Aphanomyces euteiches* during *zoosporogenesis*. Can. J. Bot. 51:413–420.
38. Homma, Y. J., J. W. Sitton, R. J. Cook, and K. M. Old. 1979. Perforation and destruction of pigmented hyphae of *Gaeumannomyces graminis* by vampyrellid amoebae from Pacific Northwest wheat field soils. Phytopathology 69:1118–1122.
39. Ioannou, N., R. W. Schneider, R. G. Grogan, and J. M. Duniway. 1977. Effect of water potential and temperature on growth, sporulation, and production of microsclerotia by *Verticillium dahliae*. Phytopathology 67:637–644.
40. Jones, F. G. W. 1975. The soil as an environment for plant parasitic nematodes. Ann. Appl. Biol. 79:113–139.
41. Keim, R. and R. K. Webster. 1974. Effect of soil moisture and temperature on viability of sclerotia of *Sclerotium oryzae*. Phytopathology 64:1499–1502.
42. Kerr, A. 1964. The influence of soil moisture on infection of peas by *Pythium ultimum*. Austr. J. Biol. Sci. 17:676–685.
43. King, C. J., and E. D. Eaton. 1934. Influence of soil moisture on longevity of cotton root-rot sclerotia. J. Agr. Res. 49:793–798.
44. Kliejunas, J. R., and W. H. Ko. 1974. Effect of motility of *Phytophthora palmivora* zoospores on disease severity in papaya seedlings and substrate colonization in soil. Phytopathology 64:426–428.
45. Kozlowski, T. T. 1978. Water deficits and plant growth. Vol. V. Water and plant disease. Academic Press, New York, San Francisco, and London.
46. Ledingham, R. J. 1970. Survival of *Cochliobolus sativus* conidia in pure culture and in natural soil at different relative humidities. Can. J. Bot. 48:1893–1896.
47. Linderman, R. G., and T. A. Toussoun. 1968. Pathogensis of *Thielaviopsis basicola* in nonsterile soil. Phytopathology 58:1,578–1,583.
48. Lipps, P. E. 1980. The influence of temperature and water potential on asexual reproduction by *Pythium* spp. associated with snow rot of wheat. Phytopathology 70:794–797.
49. ———, and ———. 1978. Snow rot of winter wheat in Washington. Phytopathology 68:1,120–1,127.
50. Lumsden, R. D., and W. A. Ayres. 1975. Influence of soil environment on the germinability of constitutively dormant oospores of *Pythium ultimum*. Phytopathology 65: 1,101–1,107.
51. MacDonald, J. D., and J. M. Duniway. 1978a. Influence of the matric and osmotic components of water potential on zoospore discharge in *Phytophthora*. Phytopathology 68: 751–757.
52. ———, and ———. 1978b. Influence of soil texture and temperature on the motility of *Phytophthora cryptogea* and *P. megasperma* zoospores. Phytopathology 68:1,627–1,630.
53. ———, and ———. 1979. Use of fluorescent antibodies to study the survival of *Phytophthora megasperma* and *P. cinnamomi* zoospores in soil. Phytopathology 69: 436–441.
54. MacNish, G. C. 1973. Survival of *Gaeumannomyces graminis* var. *tritici* in field soil stored in controlled environments. Austr. J. Biol. Sci. 26:1319–1325.
55. Manandhar, J. B., and G. W. Bruehl. 1973. *In vitro* interactions of *Fusarium* and *Verticillium* wilt fungi with water, pH, and temperature. Phytopathology 63:413–419.
56. Menzies, J. D., and G. E. Griebel. 1967. Survival and saprophytic growth of *Verticillium dahliae* in uncropped soil. Phytopathology 57:703–709.
57. Mozumder, B. K. G., N. E. Caroselli, and L. S. Albert. 1970. Influence of water activity, temperature, and their interaction on germination of *Verticillium albo-atrum* conidia. Plant Physiol. 46:347–349.
58. Old, K. M. 1977. Giant soil amoebae cause perforation of conidia of *Cochliobolus sativus*. Trans. Br. Mycol. Sco. 68:277–281.
59. Papendick, R. I., and G. S. Campbell. 1975. Water potential in the rhizosphere and plant and methods of measurement and experimental control. p. 39–49. *In* G. W. Bruehl (ed.) Biology and control of soil-borne plant pathogens. Am. Phytopathol. Soc. St. Paul, Minn.

60. ———, and R. J. Cook. 1974. Plant water stress and development of *Fusarium* foot rot in wheat subjected to different cultural practices. Phytopathology 64:358–363.

61. Pfender, W. F., R. B. Hine, and M. E. Stanghellini. 1977. Production of sporangia and release of zoospores by *Phytophthora megasperma* in soil. Phytopathology 67:657–663.

62. Reeves, R. J. 1975. Behaviour of *Phytophthora cinnamomi* Rands in different soils and water regimes. Soil Biol. Biochem. 7:19–24.

63. Schoeneweiss, D. F. 1975. Predisposition, stress, and plant disease. Annu. Rev. Phytopathol. 13:193–211.

64. Sitton, J. W., and R. J. Cook. 1980. Comparative morphology and survival ability of chlamydospores of *Fusarium roseum* 'Culmorum' and 'Graminearum.' Phytopathology 70: in press.

65. Smith, A. M. 1972. Drying and wetting sclerotia promotes biological control of *Sclerotium rolfsii* Sacc. Soil Biol. Biochem. 4:119–123.

66. Sneh, B., and D. L. McIntosh. 1974. Studies on the behavior and survival of *Phytophthora cactorum* in soil. Can. J. Bot. 52:795–802.

67. Stanghellini, M. E. 1974. Spore germination, growth and survival of *Pythium* in soil. Proc. Am. Phytopathol. Soc. 1:211–214.

68. ———, and T. J. Burr. 1973. Effect of soil water potential on disease incidence and oospore germination of *Pythium aphanidermatum*. Phytopathology 63:1496–1498.

69. ———, and J. G. Hancock. 1971. Radial extent of the bean spermosphere and its relation to the behavior of *Pythium ultimum*. Phytopathology 61:165–168.

70. Sterne, R. E., G. A. Zentmyer, and M. R. Kaufmann. 1977a. The effect of matric and osmotic potential of soil on *Phytophthora* root disease of *Persea indica*. Phytopathology 67:1491–1494.

71. ———, ———, and ———. 1977b. The influence of matric potential, soil texture, and soil amendment on root disease caused by *Phytophthora cinnamomi*. Phytopathology 67:1,495–1,500.

72. Sung, J. M., and R. J. Cook. 1981. Effect of water potential on reproduction and spore germination by *Fusarium roseum* 'Graminearum,' 'Culmorum,' and 'Avenaceum.' Phytopathology 71: in press.

73. Timian, R. G. 1959. Viability and pathogenicity of stored *Helminthosporium sorokinianum* conidia. Plant Dis. Rep. 43:1105–1107.

74. Wearing, A. H., and L. W. Burgess. 1979. Water potential and the saprophytic growth of *Fusarium roseum* 'Graminearum.' Soil Biol. Biochem. 11:661–667.

75. Westerlund, F. V., R. N. Campbell, R. G. Grogan, and J. M. Duniway. 1978. Soil factors affecting the reproduction and survival of *Olpidium brassicae* and its transmission of big vein agent to lettuce. Phytopathology 68:927–935.

76. Wilson, J. M., and D. M. Griffin. 1975. Respiration and radial growth of soil fungi at two osmotic potentials. Soil Biol. Biochem. 7:269–274.

77. Zan, K. 1962. Activity of *Phytophthora infestans* in soil in relation to tuber infection. Trans. Br. Mycol. Soc. 45:205–221.

CHAPTER 5

Water Potential as a Selective Factor in the Microbial Ecology of Soils[1]

D. M. GRIFFIN[2]

INTRODUCTION

It has long been realized that changes in the water content of soil have profound effects on microbial activity. Unfortunately, early work was greatly impeded by the absence of an adequate theoretical framework and of appropriate techniques. The data from a great volume of work cannot therefore be interpreted with any certainty (Griffin, 1963a). However, since 1960 the importance of precise analysis, centering on the concept of water potential, has been increasingly realized. Even now, many articles report work in which adequate control of the soil water regime was not obtained.

The soil water regime has many interacting components and factors which singly or in combination, directly or indirectly, exert a selective pressure on microorganisms. These include matric and osmotic or solute potentials, solute diffusion, concentrations and diffusion rates of gases and vapors, and pore size distribution (Griffin, 1972). It is now possible to devise experiments to evaluate with some confidence the significance of these selective factors in the ecology of individual species and in the incidence of plant diseases.

This chapter places emphasis on issues where confusion or ambiguity may arise and gives examples of the selective effects of the various factors associated with water potential. The literature cited is by no means complete, but sufficient to illustrate the various issues. No attempt has been made to describe in detail a series of appropriate techniques lest these be adopted uncritically as having general validity. Fungi differ so much in the details of their reaction to changes in the soil water regime that each study needs to be considered separately. As important factors are identified then appropriate techniques can be devised for a more thorough in-

[1] Contribution from the Dep. of Forestry, The Australian National Univ., Canberra.

[2] Professor of forestry.

vestigation. The literature cited incorporates techniques that have been appropriate in specific cases and may well form the basis for new experiments.

SELECTIVE EFFECTS AT HIGH WATER POTENTIALS

The analysis of changes in the soil ecosystem occurring at high water potentials, which may be taken arbitrarily to be potentials exceeding -1 bar, requires particular care because of the many factors changing simultaneously.

Soil water potential (ψ) is the sum of matric (ψ_m), osmotic or solute (ψ_π) and pressure (ψ_p) potentials, but the latter component under most conditions can be ignored. If $\psi > -1$ bar, the effects of ψ_π are negligible because of the extreme dilution. Consideration of ψ_π will be deferred to a later section. When ψ_m is high, interaction between ψ_m and pore size distribution is very important in regard to the activity of unicellular microorganisms in soil. Pore size distribution can be evaluated from the water sorption isotherm (moisture characteristic curve) and be adjusted experimentally by compaction (Kerr, 1964), by the addition of graduated amounts of sand to a clay soil or by the creation of artificial soils from inert grits of known particle size range (Griffin, 1963b; Wong and Griffin, 1976).

As matric potential falls, pores drain according to the formula $r = -1.47/\psi_m$ where r = radius of pore neck (μm) and m = matric potential (bar, where 1 bar = 100 kPa = 100 Jkg^{-1}). Matric potential through interaction with pore size distribution will therefore determine the ability of different microbial structures of varying sizes to move through water-filled pathways in soil by Brownian movement or their own intrinsic mobility.

Matric Potential and Zoospores

Zoospores of *Phytophthora cinnamomi* have been shown to move in helical pathways with an amplitude of 26 to 128 μm (Allen and Newhook, 1973). Further, frequent contact of the zoospores with solid surfaces induces encystment so that the radius of water-filled pores allowing prolonged motility will probably be at least 100 μm and even as much as 1 mm. Pore necks of this radius drain between -0.015 and -0.0015 bar ψ_m.

The validity of these suppositions has been precisely determined by Duniway (1976). Soil was placed in sintered glass Buchner funnels and the matric potential adjusted. Movement of zoospores of *Phytophthora cryptogea* was then measured. Zoospores readily swam 25 to 35 mm through a relatively coarse soil mixture when the ψ_m was > -0.015 bar. The lowest matric potential at which movement was detected was approximately -0.05 bar. In most finer-textured loams tested, there was a marked reduction in motility even at $\psi_m = -0.001$ bar, compared to that in a flooded soil. Such soils were shown to have few pores of the necessary large size so that even at high potentials, suitable water-filled pathways were few and probably very short.

The production and germination of zoosporangia are also very sensitive to matric potential. Many workers have shown that production is most abundant at very high potentials, declining rapidly when ψ_m is < -0.4 bar (Sneh and McIntosh, 1974; Duniway, 1975a, 1976b; Reeves, 1975; Ioannou and Grogan, 1977; Pfender et al., 1977). Some of these studies have shown that production of zoosporangia was poor or absent at $\psi_m = 0$, others have reported it to be abundant. It seems probable that oxygen availability is critical in such flooded soils. Mycelia near the soil-air interface, or in some surface water layers, may be sufficiently well aerated to permit sporangial formation whereas mycelia more deeply buried cannot do so.

Matric potential has been shown to be a factor in determining the mode of germination of sporangia of *Phytophthora megasperma*. Germination is by zoospores at $\psi_m = 0$, by both zoospores and germ tube at -0.05 bar and entirely by germ tube at -0.1 bar (Pfender et al., 1977). The significance of the alternative modes to infection is not known. Germination of sporangia to produce and liberate zoospores occurs most abundantly at $\psi_m > -0.006$ bar and is greatly impaired at $\psi_m = -0.1$ bar (MacDonald and Duniway, 1977; Sugar et al., 1977).

Infection of host plants by zoospores of *Phytophthora* spp. is extraordinarily sensitive to slight changes in ψ_m which acts as a selective factor at many stages in the life cycle of these fungi. Most infection probably occurs when a) ψ_m is initially between -0.025 and -0.3 bar permitting abundant sporangial production in a wet but aerated soil; b) ψ_m subsequently increases to 0 bar to ensure zoospore production and liberation; and c) soil is maintained at $\psi_m > -0.001$ bar to ensure unimpeded zoospore migration in the soil pores. Poor control of ψ_m and of pore size distribution may well inhibit one of these stages and lead to false conclusions in experiments on diseases caused by fungi (Ioannou and Grogan, 1977).

Bacterial Activity

A bacterial colony restricted to a microsite within soil will remain active only while nutrients are available to it. Continued activity involves either movement of the bacteria to new substrate or the movement of the substrate to the bacteria by plant growth, soil movement or, in some soils, by solute diffusion (Griffin, 1972). Soil water may affect bacterial activity in two major ways: by restricting movement of the bacteria themselves to new loci of nutrients; or by restricting metabolism of established colonies through nutrient deficiencies.

Bacteria experience similar, though less severe, constraints to zoospores in regard to their movement through soil by flagellar or Brownian activity. Bacterial movement becomes negligible as soils drain between about -0.2 and -1 bar ψ_m (Wong and Griffin, 1976). Many bacterial activities in soil have been shown to decrease sharply as ψ_m falls to between -0.5 and -3 bars. These include certain chemical transformations (Griffin, 1972), lysis of fungal mycelium (Sneh and McIntosh, 1974), and bacterial respiration (Wilson and Griffin, 1975). Colonization by bacteria of the lenticels of potato (*Solanum tuberosum*) tubers declines when ψ_m is < -0.1 bar and this has been shown to be correlated with invasion of the lenticels by *Streptomyces scabies* (Lewis, 1970; Lapwood and Adams, 1973) and so with the incidence of scab. It seems most probable that these declines are all caused primarily by reduced bacterial mobility.

The 'relative competitive advantage' (Baker and Cook, 1974) of bacteria and fungi in soil is thus likely to change markedly as ψ_m declines, because filamentous fungi do not suffer from the limitation of movement imposed by a unicellular structure. This interaction between fungi and bacteria will be discussed later.

Solute Diffusion

Zoospores and bacteria have been shown to be affected by even slight reduction in ψ_m below saturation. Similar changes in ψ_m can also affect solute diffusion. As a first approximation, the rate of diffusion of nonpolar molecules is proportional to the volumetric water content of the soil. In coarse soils, reduction in ψ_m below zero may therefore rapidly affect solute diffusion; even in loams, diffusion may be reduced by one-half at $\psi_m = -1$ bar. Kerr (1964) and Stanghellini and Hancock (1971) produced much evidence indicating that infection of germinating peas and beans by *Pythium ultimum* was strongly affected by the rate of diffusion of sugars liberated from these seeds. Cook and Flentje (1967) carried the evidence one step further by showing that loss of weight by pea (*Pisum sativum*) seeds during germination and the germination of chlamydospores of *Fusarium solani* f. *pisi* both decreased with soil water contents below 8.7%. Chlamydospore germination, however, could be maintained at these lower soil water contents if sucrose and ammonium sulfate were added to the soil, thus removing the dependence of the spores on pea exudates.

Recently, chlamydospore germination and germ tube development in *Phytophthora cinnamomi* and associated disease incidence in *Persea indica* were shown to be high in a coarse sandy loam at $\psi_m > -0.1$ bar but very low at $\psi_m = -0.25$ bar. In a clay, however, this range of potential had little differential effect. Addition of glucose and asparagine removed the differential effect of ψ_m in the sandy loam (Sterne et al., 1977). Taken together, these data strongly suggest that nutrient availability is a limiting factor for activity by this fungus at $\psi_m < -0.25$ bar in coarse-textured soils.

Aeration

Within the same range of matric potentials as are critical for bacterial and zoospore movement and for solute diffusion, soil aeration is greatly affected and the effects of altered concentrations and diffusion rates of gases need to be distinguished from those of the other factors. A detailed consideration of the roles of oxygen and carbon dioxide falls outside the scope of this symposium and has been reviewed elsewhere (Griffin, 1968, 1972). The soil system is so complex that it is rarely possible to demonstrate conclusively that a shortage of oxygen is the limiting factor for a given microbial activity. Reduction in ψ_m will soon cause drainage of the major pore systems lying between aggregates but the aggregates themselves will frequently remain saturated. Thus, aeration will have improved for a spore or hypha lying on the edge of a major pore whereas this will probably not be so within aggregates. In saturated spherical soil crumbs, all but the outer shell of about 3-mm thickness is likely to be anaerobic depending on such things as the rate of respiration and temperature. In detail, therefore, the effects of reduced ψ_m acting on a microorganism through changes in aeration will depend upon how the population of that organism is distributed on and within soil aggregates. Such information is rarely available.

Generalizations can be made, however. Thus, the respiratory quotient for a number of soils dropped from greater than two to unity, as the gas-filled pore space increased from 10 to 20% (Rixon and Bridge, 1968). In most soils the respiratory quotient is not affected until the matric potential decreases to about -0.1 bar. Such a change to predominantly aerobic respiration must signify great changes in the active microbial population and the probable commencement of fungal activity at an intensity comparable with that of bacteria. In such a situation it is extremely difficult to distinguish the general antagonistic effect of bacteria against fungi from the deleteriour effects on fungi of low oxygen concentrations caused by bacterial metabolism.

The role of carbon dioxide is even more difficult to evaluate, either experimentally or theoretically, because of its solubility in water and reactivity in the soil (Griffin, 1972). Unless the conditions are quite exceptional, however, such as in flooded fields, data indicate that carbon dioxide concentrations are unlikely to be significant in soil microbiology.

Detailed analysis of the roles of water potential, oxygen, and carbon dioxide in the ecology of *Armillariella elegans* (Smith and Griffin, 1971) and *Verticillium dahliae* (Ioannou et al., 1977a, 1977b, 1977c) have been made and may provide useful techniques for studying these parameters. Biologically active gasses in soil are not restricted to oxygen and carbon dioxide. For example, the effect of ethylene on microbial activity has been studied and would appear to have some significant implications (Balis, 1976; Smith, 1976). Cook and Smith (1977) reported that the production of ethylene in soil is very sensitive to changes in matric potential. Little is known in precise terms of the effects of water potential on the production and diffusion of the wide range of other gases and volatile chemicals that may be found in soil.

Bulk density

Bulk density is equal to the mass of solids in a sample divided by the total volume of solids, liquids, and gases in that sample. It is rarely measured in soil microbiology experiments yet is of considerable significance. An increase in bulk density at constant gravimetric water content increases volumetric water content and hence solute diffusion. Simultaneously, however, gaseous diffusion is reduced and pore size distribution altered, frequently with loss of the larger pores. Infection of pea seeds by *Pythium ultimum* increased with bulk density, almost certainly because of enhanced diffusion of exuded carbohydrates (Kerr, 1964).

SELECTIVE EFFECTS OF MODERATE WATER POTENTIALS

Within the range of -1 to -15 bars ψ_m, relative competitive advantage is still of great significance. It may best be illustrated by the rather general case of the alteration in the relative contribution of bacteria and fungi, as groups, to overall soil respiration. As ψ_m declined to between -3 and -6 bars, bacterial respiration fell rapidly and at -20 bars was negligible (Wilson and Griffin, 1975). Total microbial respiration, however, showed little decline between -8 and -30 bars ψ_m and most of this was probably attributable to fungi. Other data support such an interpretation (Anderson and Domsch, 1973; Faegri et al., 1977).

In general, bacterial activity continues to decline at matric potentials below those where any marked effect on motility is likely and becomes negligible for most purposes near -15 bars (Griffin, 1972). The factors involved here cannot be specified with any certainty but the decline is probably not attributable to a direct effect of potential on respiration and growth. In pure culture, these activities continue, though at slow rates, for many species at osmotic potentials as low as -100 bars.

As ψ_m falls, the relative competitive advantage between bacteria and fungi moves progressively towards the latter, and this may well be a major factor in the increased incidence of fungal disease. An excellent example of this concerns the disease of wheat (*Triticum aestivum*) caused by *Fusarium culmorum*. At potentials greater than -8 to -10 bars, chlamydospore germ tubes either rapidly lysed or reformed chlamydospores (Cook and Papendick, 1970), and it was only at potentials lower than this critical range that continued mycelial growth occurred. Such growth was correlated with greatly reduced bacterial population.

These data strongly suggest that *F. culmorum* can withstand bacterial antagonism only in relatively dry soils, a hypothesis supported by the fact that incorporation of neomycin and streptomycin eliminated germ tube lysis at -0.9 bar. A comparable situation has been described for infection of pea epicotyls by *F. solani* f. *pisi*. In neither case can a direct effect of potential on the host be ruled out in determining disease incidence, but altered relative compatitive advantage of bacteria and pathogen seems the most likely factor.

Due to this strong interaction between ψ_m and the relative importance of fungal and bacterial activity, data derived from pure culture experiments are of only limited value in aiding the understanding of what actually happens in soil. In the absence of bacterial antagonism, for instance, a fungus may well respond to changes in ψ_m in quite different ways to those when bacteria are present.

At moderate soil water potentials the contribution of the osmotic component (ψ_π) may have microbiological significance, at least where zoospores are involved (Hoch and Mitchell, 1973). The effects of ψ_m and

ψ_π on microbial growth and metabolism have been discussed elsewhere (Griffin, 1978; Harris, 1980). It suffices here to note that fungi are frequently, but not always, more tolerant to low ψ_π than of low ψ_m. It is suspected that the difference is largely due to alteration in such factors as solute diffusion, that are linked with matric potential. Further, the optimum for vegetative growth of many fungi which occurs at an osmotic potential of approximately -8 bars, both in agar culture and in natural substrates in soil, is not found when the potential is controlled matrically (Cook et al., 1972). It follows that, although in a narrow thermodynamic sense, matric and osmotic potentials are additive and interchangeable in their effects, this may not be so in field situations. A situation where $\psi_m = -13$ bars and $\psi_\pi = -2$ bars will be far more restricting to microbial activity than where $\psi_m = -2$ bars and $\psi_\pi = -13$ bars. When using techniques such as isopiestic equilibration or the thermocouple psychrometer which measure water potential, i.e. $\psi_m + \psi_\pi$, it is therefore advisable to determine ψ_π separately if it is thought to be a major component. Otherwise comparisons between systems of identical water potentials may be fallacious because of differences in the ψ_m and ψ_π components.

In experimental systems where the osmotic potential is to be adjusted, it is of course necessary to use a solute whose effects are predominantly upon ψ_π. Many salts are toxic, carbohydrates may be used as an additional energy source and high molecular polyethylene glycols restrict gaseous diffusion (Mexal et al., 1975). Many experiments have been conducted using potassium chloride as the added solute and results indicate that its main effect is to facilitate a change in ψ_π. This salt is known to be important in osmoregulation (Hellebust, 1976) and it is probably the solute of first choice in experiments on osmotic potential.

Frequently, ψ_π is adjusted in agar gels but the effect of the gel itself on ψ_π and ψ_m must be considered (Gardner et al., 1972; Sterne et al., 1976). To prevent extraneous effects, the agar content of the gel should be kept to the lowest practicable level.

SELECTIVE EFFECTS AT LOW WATER POTENTIALS

In pure culture, the growth of some fungi is negligible at -20 bars whereas others are able to grow at -400 bars. Clearly, there is ample opportunity for water potential per se to act as a selective factor when ψ is < -15 bars. Often, however, activity is determined not only by water potential but also by microbial antagonism.

There is now considerable evidence that interspecific antagonism between fungi in soil limits the activity of most species to water potentials higher than those which they can tolerate in pure culture. A few fungi, mainly xerophytic species in the genera *Aspergillus* and *Penicillium* and perhaps some of the *Fusarium* spp., grow almost equally well in pure and mixed cultures at low water potentials, no doubt due to the greatly reduced antagonism from the general microflora even in the latter. Using isopiestic techniques to establish and maintain stable water potentials, Chen and Griffin (1966) studied the colonization of hair fragments lying on soil and clearly demonstrated the importance of low water potential as a selective ecological factor.

At these lower potentials, antagonism between fungi is probably of greatest significance but that between fungi and actinomycetes may still be important. Thus, antibiotic production by a *Streptomyces* sp. and the antagonistic effect on three test fungi were maximal at ψ_π between −20 and −35 bars (Wong and Griffin, 1974). This range of ψ_π also allowed maximum production of antibiotic by *Cephalosporium gramineum* (Bruehl et al., 1972) so the action of antibiotics may be of particular importance at these lower potentials.

When considering water potential as a limiting factor, Cook and Papendick (1972) have stressed the need to take into account its heterogeneous distribution within a field soil. In surface soils, ψ may be less than −100 bars, while that in the rhizosphere at greater depth may be only −30 bars. In such situations, fungi will often be exposed to potentials much less than that of the bulk soil. Such potentials are too low for the growth of the wheat pathogen *Gaeumannomyces graminis* but are well-suited to *Fusarium culmorum* (Cook et al., 1972). The selective effect of water potential therefore probably accounts for the higher disease incidence by the former species under irrigated conditions in the Pacific northwest of the USA, and by the latter pathogen under dryland conditions. Such correlations between in vitro growth patterns and pathogenicity when ψ is the variable could not be detected, however, for *Fusarium oxysporum* f. sp. *vasinfectum* and *Verticillium alboatrum* (Manandhar and Bruehl, 1973).

At low water potentials, and to a lesser degree at higher potentials, there are important interactions between ψ and temperature and nutrition that determine fungal response (Sterne et al., 1976; Griffin, 1978). Thus, growth is possible at lower potentials if the substrate has high concentrations of nutrients. This must be considered when extrapolating from data obtained from growth on nutrient agars to growth in soil.

There are many reports of a selective effect of ψ_π in saline habitats (Tresner and Hayes, 1971; Davidson, 1974; Bryne and Jones, 1975; Harrison and Jones, 1975) but most of these effects are probably attributable to a general response to ψ rather than to ψ_π specifically. Cook et al. (1972) reported that at most potentials there was little difference between the effects of ψ_m and ψ_π on the growth of *Fusarium culmorum* on straw buried in soil. In the surface layers of saline soils, ψ_π may well be less than −100 bars even though ψ_m is approximaely −15 bars (Cook and Papendick, 1972). Little attention has been given to the effects on microorganisms in situ in soil where the ψ_π is within the range occurring in heavily fertilized or saline soils.

CONCLUSIONS

In the past, the soil water regime was conventionally expressed solely in values associated with gravimetric soil water content and the deficiencies of this situation are now generally appreciated. Techniques are readily available to control matric and solute potentials, and their sum, and these are more regularly being evaluated. It is important, however, that soil microbiologists should not be lulled into a false sense of security concerning the use of water potential. It is true that measurements of water potential convey far more useful information than do those of gravimetric water content, yet either one alone leaves much unevaluated. Pore size distribution, aeration, bulk density, and volumetric water content, are all interrelated, and may profoundly affect the course of microbial events in soil.

Experiments are most easily conducted in pure culture, with ψ_π rather than ψ_m as the variable, and often at 25 C with high nutrient concentrations. Such experiments with fungi provide important data yet neglect significant interactions, particularly those between fungi and bacteria, and underrate the effects of ψ_m in nutrient deficient sites at temperatures below 25 C. As discussed earlier, the response of bacteria to ψ_π is no guide whatsoever to their response to ψ_m.

LITERATURE CITED

1. Allen, R. N., and F. J. Newhook. 1973. Chemotaxis of zoospores of *Phytophthora cinnamomi* to ethanol in capillaries of spore dimensions. Trans. Br. Mycol. Sco. 61:287–302.
2. Anderson, J. P. E., and K. H. Domsch. 1973. Quantification of bacterial and fungal contributions to soil respiration. Archiv. Mikrobiol. 93:113–127.
3. Baker, K. R., and R. J. Cook. 1974. Biological control of plant pathogens. W. H. Freeman and Co., San Francisco.
4. Balis, C. 1976. Ethylene-induced volatile inhibitors causing soil fungistasis. Nature. 2509:112–114.
5. Bruehl, G. W., B. Cunfer, and M. Toivianinen. 1972. Influence of water potential on growth, antibiotic production and survival of *Cephalosporium gramineum*. Can. J. Plant Sci. 52:417–423.
6. Byrne, P., and E. B. Gareth Jones. 1975. Effect of salinity on spore germination of terrestrial and marine fungi. Trans. Br. Mycol. Soc. 64:497–503.
7. Chen, A. W., and D. M. Griffin. 1966. Soil physical factors and the ecology of fungi. V. Further studies in relatively dry soils. Trans. Br. Mycol. Soc. 49:419–426.
8. Cook, R. J., and N. T. Flentje. 1967. Chlamydospore germination and germling survival of *Fusarium solani* f. *pisi* in soil as affected by soil water and pea seed exudation. Phytopathology. 57:178–182.
9. ————, and R. I. Papendick. 1970. Soil water potential as a factor in the ecology of *Fusarium roseum* f. sp. *cerealis* 'Culmorum'. Plant Soil 32:131–145.
10. ————, and ————. 1972. Influence of water potential of soils and plants on root disease. Annu. Rev. Phytopathol. 10:349–374.
11. ————, ————, and D. M. Griffin. 1972. Growth of two root rot fungi as affected by osmotic and matric water potentials. Soil Sci. Soc. Am. Proc. 36:78–82.
12. ————, and A. M. Smith. 1977. Influence of water potential on production of ethylene in soil. Can. J. Microbiol. 23:811–817.
13. Davidson, D. E. 1974. Salinity tolerance and ecological aspects of some microfungi from saline and non-saline soils in Wyoming. Mycopathol. Mycol. Appl. 54:181–188.
14. Duniway, J. M. 1975a. Formation of sporangia by *Phytophthora drechsleri* in soil at high matric potentials. Can. J. Bot. 53:1270–1275.
15. ————. 1975b. Limiting influence of low water potential on the formation of sporangia by *Phytophthora drechsleri* in soil. Phytopathology 65:1089–1093.
16. ————. 1976. Movement of zoospores of *Phytophthora cryptogea* in soils of various textures and matric potentials. Phytopathology 66:877–882.

17. Faegri, A., V. L. Torsvik, and J. Goksoyr. 1977. Bacterial and fungal activities, in soil: Separation of bacteria and fungi by a rapid fractionated centrifugation technique. Soil Biol. Biochem. 9:105–112.

18. Gardner, W. R., F. N. Dalton, and R. F. Harris. 1972. Thermocouple psychrometry for the study of water relations of soil microorganisms. p. 150–153. *In* R. W. Brown and B. P. van Haveren (ed.) Psychrometry in water relations research. Utah Agric. Exp. Stn. Logan, Utah.

19. Griffin, D. M. 1963a. Soil moisture and the ecology of soil fungi. Biol. Rev. Cambridge Philos. Soc. 38:141–166.

20. ————. 1963b. Soil physical factors and the ecology of fungi: I. Behaviour of *Curvularia ramosa* at small soil water suctions. Trans. Br. Mycol. Soc. 46:273–280.

21. ————. 1968. A theoretical study relating the concentration and diffusion of oxygen to the biology of organisms in soil. New Phytologist 67:561–577.

22. ————. 1972. Ecology of soil fungi. Chapman and Hall. London.

23. ————. 1978. Effect of soil moisture on survival and spread of pathogens. p. 175–197. *In* T. T. Kozlowski (ed.) Water deficits and plant growth. Vol. 5, Academic Press, New York.

24. Harris, R. F. 1980. Effect of water potential on microbial growth and metabolism. In press. *In* J. F. Parr and W. R. Gardner (ed.) The concept of water potential applied to soil microbiology and biochemistry. Am. Soc. of Agron., Madison, Wis.

25. Harrison, J. L., an E. B. Gareth Jones. 1975. The effect of salinity on sexual and asexual sporulation of members of the *Saprolegniaceae*. Trans. Br. Mycrol. Soc. 65:389–394.

26. Hellebust, J. A. 1976. Osmoregulation. Annu. Rev. Plant Physiol. 27:485–505.

27. Hoch, H. C., and J. E. Mitchell. 1973. The effects of osmotic water potentials on *Aphanomyces euteiches* during zoosporogenesis. Can. J. Bot. 51:413–420.

28. Ioannou, N., and R. G. Grogan. 1977. The influence of soil matric potential on the production of sporangia by *Phytophthora parasitica* in relation to its isolation from soil by baiting techniques. Am. Phytopathol. Soc. Proc. 4:173.

29. ————, R. W. Schneider, and R. G. Grogan. 1977a. Effect of oxygen, carbon dioxide, and ethylene on growth, sporulation, and production of microsclerotia by *Verticillium dahliae*. Phytopathology 67:645–650.

30. ————, ————, and R. G. Grogan. 1977b. Effect of flooding on the soil gas composition and the production of microsclerotia by *Verticillium dahliae* in the field. Phytopathology. 67:651–656.

31. ————, ————, ————, and J. M. Duniway. 1977c. Effect of water potential and temperature on growth, sporulation, and production of microsclerotia by *Verticillium dahliae*. Phytopathology 67:637–644.

32. Kerr, A. 1964. The influence of soil moisture on infection of peas by *Pythium ultimum*. Austr. J. Biol. Sci. 17:676–685.

33. Lapwood, D. H., and M. J. Adams. 1973. The effect of a few days of rain on the distribution of common scab (*Streptomyces scabies*) on young potato tubers. Ann. Appl. Biol. 73:277–283.

34. Lewis, B. G. 1970. Effects of water potential on the infection of potato tubers by *Streptomyces scabies* in soil. Ann. Appl. Biol. 66:83–88.

35. MacDonald, J. D., and J. M. Duniway. 1977. Influence of the matric and osmotic components of water potential on zoospore discharge in *Phytophthora*. Am. Phytopathol. Soc. Proc. 4:147.

36. Manandhar, J. B., and G. W. Bruehl. 1973. In vitro interactions of *Fusarium* and *Verticillium* fungi with water, pH, and temperature. Phytopathology 63:413–418.

37. Mexal, J., J. T. Fisher, J. Osteryoung, and C. P. P. Reid. 1975. Oxygen availability in polyethylene glycol solutions and its implications in plant-water relations. Plant Physiol. 55:20–24.

38. Pfender, W. F., R. B. Hine, and M. E. Stanghellini. 1977. Production of sporangia and release of zoospores by *Phytophthora megasperma* in soil. Phytopathology 67:657–663.

39. Reeves, R. J. 1975. Behaviour of *Phytophthora cinnamomi* Rands. in different soils and water regimes. Soil Biol. Biochem. 7:19–24.

40. Rixon, A. J., and B. J. Bridge. 1968. Respiratory quotient arising from microbial activity in relation to matric suction and air-filled pore space of soil. Nature. 218:961–962.

41. Smith, A. M. 1976. Ethylene in soil biology. Annu. Rev. Phytopathol. 14:53–73.

42. ————, and D. M. Griffin. 1971. Oxygen and the ecology of *Armillariella elegans* Heim. Austr. J. Biol. Sci. 24:231–262.

43. Sneh, B., and D. L. McIntosh. 1974. Studies on the behavior and survival of *Phytophthora cactorum* in soil. Can. J. Bot. 52:795–802.

44. Stanghellini, M. E., and J. G. Hancock. 1971. Radial extent of the bean spermosphere and its relation to the behaviour of *Pythium ultimum*. Phytopathology 61:165–168.

45. Sterne, R. E., G. A. Zentmyer, and F. T. Bingham. 1976. The effect of osmotic potential and specific ions on growth of *Phytophthora cinnamomi*. Phytopathology 66: 1,398–1,402.

46. ————, ————, and M. R. Kaufmann. 1977. The influence of matric potential, soil texture, and soil amendment on root disease caused by *Phytophthora cinnamomi*. Phytopathology. 67:1,495–1,500.

47. Sugar, D., S. M. Mircetich, and J. M. Duniway. 1977. Influence of soil matric potential on the formation and germination of sporangia by *Phytophthora* spp. causing root and crown rot of cherry trees. Am. Phytopathol. Soc. Proc. 4:211.

48. Tresner, H. D., and J. A. Hayes. 1971. Sodium chloride tolerance of terrestrial fungi. Appl. Microbiol. 22:210–213.

49. Wilson, J. M., and D. M. Griffin. 1975. Water potential and the respiration of microorganisms in the soil. Soil Biol. Biochem. 7:199–204.

50. Wong, P. T. W., and D. M. Griffin. 1974. Effect of osmotic potentials on streptomycete growth, antibiotic production, and antagonism to fungi. Soil Biol. Biochem. 6:319–325.

51. ————, and ————. 1976. Bacterial movement at high matric potentials. I. In artificial and natural soils. Soil Biol. Biochem. 8:215–218.